D1215294

Chicago Public Library
C
P
L
REFERENCE
Form 178 rev. 11-00

QUALITY ASSESSMENT of WATER and WASTEWATER

QUALITY ASSESSMENT of WATER and WASTEWATER

Mamta Tomar

Acquiring Editor:	Bob Hauserman
Project Editor:	Sylvia Wood
Marketing Manager:	Arline Massey
Cover design:	Dawn Boyd
Manufacturing Manager:	Carol Slatter
Compositor:	James Yanchak

Library of Congress Cataloging-in-Publication Data

Tomar, Mamta
Quality assessment of water and wastewater / Mamta Tomar
p. cm.
Includes bibliographical references and index.
ISBN 1-56670-382-4
1. Water--Analysis--Laboratory manuals. 2. Sewage--Analysis--Laboratory manuals. I. Title.
[DNLM: 1. Hepatitis B virus. QW 710 G289h]
TD380.T66 1999
628.1'61--dc21
DNLM/DLC
for Library of Congress

99-10114
CIP

International Standard Book Number 1-56670-382-4
Library of Congress Card Number 99-10114
Printed in the United States of America 1 2 3 4 5 6 7 8 9 0
Printed on acid-free paper

Preface

Water is as essential for human and all other living beings as food. And, not only for drinking, humans need water for various other purposes like bathing, washing, cooking, industrial, agricultural, and recreational activities. Thus, the availability of adequate water supply in terms of its quality and quantity is essential for the existence of life. It should be free from pathogens – disease-causing microbes and toxic or physiologically undesirable chemicals or biological materials.

Water is available in nature in the form of ground and surface waters, i.e., rivers, lakes, ponds, and oceans. Though self-purification mechanisms like physical, chemical, and microbiological processes of natural water bodies are carried out in nature, their waters are very rarely suitable for direct consumption by humans. Disposal of human/animal wastes and wastes generated from industrial, commercial, and agricultural establishments releases a variety of inorganic, organic, and biological contaminants to natural water bodies. Hence, natural waters require physical, chemical, and biological treatment, depending on the nature of existing pollutants, before being supplied for domestic use.

To plan and implement the type and extent of treatment, natural waters must be analyzed for physical, chemical, and microbiological parameters. After appropriate treatment in a water treatment plant, the quality of water is again tested to ensure its suitability for human consumption. The suitability of water is judged on the basis of modern drinking water standards set up by different governmental and international water bodies and health agencies.

In modern and civilized societies, proper management of wastewater is a necessity, not an option. Water, the universal solvent, is largely used to collect domestic and industrial wastes and refuse in all advanced countries. The water thus produced is called wastewater. It is generally classified as domestic/municipal or industrial wastewater, depending on its source. The nature and quality of domestic wastewater differs from industrial wastewater, and the characteristics of industrial wastewater especially, vary greatly from industry to industry depending mainly on the type of production unit of the industry. Thus, on one hand, water is a convenient source for waste collection but, on the other, wastewater is an extensive source of pollution if disposed untreated into any natural water body such as a river, lake, or ocean. Wastewater contains significant amounts of suspended solids, biodegradable organics, and pathogens that would pollute the natural water and make it unsuitable for human consumption. Also, the microbial decomposition of solids available in wastewater generate malodorous and health-hazardous gaseous components that pollute the adjacent environment. For these reasons, domestic and industrial wastewaters require extensive treatment before they can be disposed of or safely reused. Different countries and international agencies have laid down limits for the quality of wastewater to be disposed of on land and in natural water bodies, as well as for the reuse of treated wastewater. In all cases, accurate and extensive testing of wastewater is essential to plan the nature and extent of treatment processes necessary to achieve a quality of wastewater that will render it acceptable for safe disposal or reuse.

To assess the level of contamination and type of treatment required, both water and wastewater require proper and reliable analytical measurements before treatment. After treatment, both must have further laboratory testing and analysis to confirm their suitability for different uses. Hence, accurate laboratory testing and analysis of water and wastewater quality form the backbone of water and wastewater technology and quality management.

The collection, treatment, and disposal of wastewater are not only the concerns of municipal/civil engineers or scientists but of environmentalists too. Its insufficient treatment or improper disposal can vastly affect and even damage the surrounding environment. Therefore, environmental pollution control agencies also monitor and regulate collection, treatment, and disposal of wastewater along with solids and gaseous wastes. Thus, environmental pollution control is closely related with water and wastewater management and technology. With the involvement of environmentalists in this field, people with varied educational backgrounds in such areas as chemical/biochemical and public health engineering, general and industrial chemistry, biochemistry, general biology, and microbiology are attracted to work in the field of water and wastewater management..

Basically, this field involves the fundamental principles of three individual streams of science, namely chemistry/biochemistry, microbiology and engineering. Microbiology is essential to understanding the nature and physiology of useful microbes carrying the assimilation of waste materials and their removal. It also helps to know the type, growth, and activity of harmful disease-causing microbes. Chemistry plays an important role in understanding the quality and quantity of different inorganic and organic pollutants in order to assess the extent of pollution in water. Finally, engineering has its own importance because engineers are the technical personnel who plan a strategy to design, operate, and maintain the treatment processes to produce water suitable for human consumption and safe disposal or reuse. Therefore, a team comprising microbiologists\biologists, chemists/biochemists, and engineers is an ideal team of professionals to deal with water and wastewater quality management and environmental pollution control problems. Individual professionals from any one discipline cannot handle them appropriately.

Engineers and chemists usually have little or no training in microbiological fundamental techniques and biochemical processes. On the other hand, biologists and microbiologists have a limited background in chemistry's basic principles. Hence, it is limiting for personnel belonging to any of the individual disciplines of science to handle both chemical and microbiological estimations with required accuracy. This manual is intended to attempt to bridge this gap. It fulfills the following objectives:

a. Provides a background of microbiology and chemistry to water and wastewater operators who have limited knowledge in both areas. Mostly the training programs of operators concentrate on treatment procedures, plant operation and management, and troubleshooting but involve very little practice in laboratory estimations. This book will fill that void.

b. Provides a basic knowledge of chemical and microbiological processes and techniques essential for analysis of water and wastewater to personnel who train operators for successful operation and monitoring of water and wastewater treatment plants.

c. Provides assistance to chemists in environmental testing laboratories who perform microbiological testing on occasions.

d. Provides supplementary information for a variety of job-related training programs in water and wastewater technology and management and refresher courses in the environmental field.

e. Provides a basic qualitative and quantitative analytical background to undergraduate students involved in water and wastewater technology and management as well as in environmental studies.

f. Brings readers having varied educational backgrounds to a common level at which all will possess similar basic knowledge to follow the chemical, biochemical, and microbiological techniques covered in this manual.

g. Prepares a basic platform for readers to grasp and perform the more sophisticated techniques for qualitative analysis of water and wastewater.

Thus, in planning this text, my basic objective was to develop the subject from the basic principles of chemistry, biochemistry, and microbiology and make it beneficial for personnel with varied educational backgrounds. The description of material has the following positive approaches:

a. All the testing procedures have been described from the beginnings of chemical and microbiological laboratory preparations, precautions, safety rules, etc.

b. The book introduces a unique approach to ascertaining the quality and contamination level of water from a wide range of sources such as ground, surface, potable water supply, marine, beaches, swimming pools, and other recreational facilities, domestic and industrial establishments. Emphasis has been placed on the variation in the characteristics of waters depending on their sources and uses. Essential test requirements for the evaluation and assessment of their quality and contamination levels are highlighted to help analysts, technicians, and operators plan the methodology for quantitative analysis of different types of water and a strategy for their treatment and utilization in various activities. Chapter 4 of this manual emphasizes the mentioned topics.

c. Management and technical personnel involved in water and wastewater quality management and pollution control need the water quality requirements and standards set up by international governmental agencies to interpret their data. These recommendations and permissible limits help them to ascertain the suitability of water for public consumption in different activities. For the first time, water quality requirements and standards followed on different continents are compiled as a ready reference for students, technicians, analysts, and plant operators working in these fields worldwide.

d. All the major inorganic and organic pollutants falling under the categories of physical and chemical estimations are grouped according to their significance in water and wastewater quality management. Their impact on the environment and health, along with the basic chemistry of their analysis and detailed methodology, are explained in an easy and systematic manner.

e. The section on microbiology provides comprehensive knowledge of the fundamentals involved in the microbiological testing of water and wastewater. It describes the nature, growth, activity, and disposal of indicator microbes linked with water contamination. The sample preservation and preparation, sample serial dilution technique, media preparation, and aseptic inoculation techniques are explained with the help of relevant figures. The significance of nonpathogenic and pathogenic microorganisms related to water pollution is also included. The explanation and methodology followed in this chapter are so clear that a pure chemist with no background in microbiology can work out microbiological estimations accurately.

f. The analytical procedures described in this manual are commonly used in water and wastewater testing laboratories and recommended as standard analytical methods for water and wastewater examination. These are very simple and well-established methods, so a moderately equipped laboratory can perform them with desired accuracy.

All the procedures included in this manual have been thoroughly studied by the author during her long analytical experience in the field of water and wastewater quality management and environmental pollution control.

AUTHOR

Mamta Tomar is a Microbiology Specialist and Advisor in SEA (the Section of Environmental Affairs), Sanitary Engineering Department of Ministry of Public Works, Kuwait. She received a Ph.D Degree in Microbial Chemistry from National Sugar Institute, Kanpur, India and her master's in Organic Chemistry from the University of Kanpur. In her long career she has worked as an advisor in different public and private establishments.

Her principal research interests are in the area of odor control in sewerage network and wastewater treatment plants by chemical and microbiological means, monitoring of water and wastewater quality, safe utilization and disposal of treated effluent and sewage sludge.

Dr. Tomar has authored and coauthored numerous technical publications. She is an active member of the New York Academy of Sciences and the Water Environment Federation, USA.

ACKNOWLEDGMENTS

The development of an investigative work requires the cooperation of colleagues, encouragement from administration, and love and support of family.

I am grateful for valuable comments and suggestions given by Mrs. Tamama H.A. Abdullah, head of the Section of Environmental Affairs (SEA), Ministry of Public Works, Kuwait. Special gratitude is extended to Professor Malcolm McCormick, Associate Dean of Royal Melbourne Institute of Technology (RMIT) and to Dr. Berry J. Meehan, Senior Lecturer, RMIT, Australia for their efforts, cooperation, and suggestions in improving the quality and presentation of this book. I am grateful to Professor P.K. Agarwal, Chairman of Biochemistry Division, National Sugar Institute, India for his valuable guidance in writing this text. I am thankful to Mohammed Hafeez, Chemist in SEA Laboratory, for his support and cooperation in establishing the accuracy of analytical procedures described in this book.

I extend special thanks to Engineer Hussein Malik, Consultant, Sanitary Engineering Department of Ministry of Public Works, Kuwait for his appreciation and encouragement throughout my working career in the Ministry.

A specific acknowledgement and grateful appreciation goes to my husband, Jayvir Singh Tomar, a Sanitary Design Engineer, for helping in organizing the text, proofreading and finalizing the index. He has devoted many days and tired eyes to help me in producing this book. My grateful appreciation goes to my loving children, Sumita and Sachin, for their expert typing of the manuscript.

Finally, I would like to thank the editorial and production staffs of CRC Press.

Contents

List of Tables

Dedication

To the memory of R.P. Bhatia,
my dear father, guide, and inspiration

Chapter 1

Introduction

Water is the most basic need of mankind. The utilization of water for various activities depends on its physical, chemical, and biological characteristics. This includes public water supply for drinking and domestic purposes, industrial activities, propagation of aquatic and wildlife, water-based recreational activities, commercial fishing, and esthetic enjoyment. And, since drinking water is considered to be the most essential use of water for life, it must be free of health hazards such as pathogens, toxins, and carcinogens.

Absolutely pure water is not available in nature because surface water absorbs particulates, carbon dioxide and other gases, and mixes with silt and inorganic matters from the environment. The problem becomes more complex when treated and untreated domestic and industrial wastes are discharged into natural water bodies. These wastes contain different organic and inorganic pollutants and various pathogenic and nonpathogenic microbes.

Thus, human waste, drinking water, and communicable diseases have a direct relation. The communicable diseases mainly transmitted by water include bacterial, viral, and protozoal infections. For this reason, drinking water must be free of any kind of microbial, organic, and inorganic pollutant. It must be safe to drink, esthetically pleasing to the taste, and suitable for domestic activities. Common diseases caused by enteric microorganisms, i.e., organisms that multiply in the human intestine and are excreted in feces, are given in Table 1.1.

The extent of contamination in water is assessed by the level of pollutants present in water. This could be obtained by regular analytical estimation of wastewater samples. The analytical examination of the quality of water and wastewater falls into three main categories.

1. Physical characteristics
2. Chemical characteristics
3. Biological characteristics

TABLE 1.1
Common Enteric Pathogenic Microorganisms, Related Diseases, and Symptoms

Groups	Microorganisms	Disease	Symptoms
Bacteria	Escherichia coli	Gastroenteritis	Acute diarrhea
	Vibrio cholerae	Cholera	Acute diarrhea & dehydration
	Salmonella typhi	Typhoid fever	High fever, diarrhea, ulcer in small intestine
	Other sps. of salmonella	Salmonellosis Gastroenteritis	Food poisoning Diarrhea
	Shigella dysenteriae	Shigellosis	Bacillary dysentery
	Leptospira	Jaundice	High fever, yellow coloration of eyes & body
	Legionella pneumophila	Leginellosis	Respiratory ailment
Viruses	Enteroviruses (polio, coxsackie & echo)	Poliomyelitis, meningitis, paralysis	Inflammation of spinal cord or bone marrow, high fever, stiffness of neck and joints
	Adenovirus	Respiratory infection & Gastroenteritis	Respiratory ailment & diarrhea
	Rheovirus	Gastroenteritis	Diarrhea
	Hepatitis A	Infectious hepatitis	Jaundice, high fever, may also affect kidneys
Protozoa	Giardia lamblia	Giardiasis	Diarrhea, nausea
	Balantidium coli	Balantidiasis	Dysentery, gastrointestinal infection
	Entamoeba histolytica	Amoebiasis	Amoebic dysentery, infection of liver & small intestine
	Isopora beli & Isopora hominis	Gastroenteritis	Gastrointestinal infection
	Cryptosporidium	Cryptosporidiosis	Diarrhea

Physical characteristics. Physical characteristics are those that reflect the palatability or esthetic acceptability of water for drinking and other domestic activities. They respond to the sense of sight, touch, taste, or smell, thus indicating the physical state of water. Color, temperature, taste, odor, turbidity, conductivity, and suspended solids are included in this category.

Chemical characteristics. Water is a universal solvent and so dissolves most naturally occurring substances as well as those produced by human activities. These substances are found at a wide range of concentrations in water depending on their abundance, solubility, and other physiochemical qualities. Thus, chemical parameters are related to the solvent capabilities of water. Alkalinity, acidity, hardness, metallic and nonmetallic, biodegradable and nonbiodegradable organics, and nutrients are chemical parameters of concern in water and wastewater quality management.

Biological characteristics. This category includes the microbiological estimation of nonpathogenic and pathogenic microbes in water. It is also related to the oxygen demand created by the ecosystem present in water and wastewater. This factor is measured as Biological Oxygen Demand (BOD). Various species of anaerobic and aerobic bacteria, fungi, and viruses fall into this category.

A well-equipped laboratory with modern amenities is essential to achieve this goal.

This manual emphasizes the following major topics:

- Different types of water and their utilization in various activities, depending on their physical, chemical, and biological characteristics
- Water-quality requirements and criteria set up by international and governmental agencies in different continents for drinking, recreational, agricultural, domestic, and commercial activities
- The simple estimation procedures and significance of each analytical test in terms of water quality and pollution assessment

Thus, this book will facilitate the work of personnel involved in the analytical estimations of water and wastewater, as well as the efforts of environmentalists, and of water and wastewater treatment plant operators in the proper operation and monitoring of the treatment system. It will hopefully help them keep our environment free of water-borne pollution, which will aid in the protection of mankind from water-borne diseases.

Chapter 2

Safety in the Laboratory

2.1 Basic Rules

Since chemicals and microbes may produce physical injury and human diseases, some basic rules should be followed in laboratories dealing with chemical and microbiological examinations of environmentally hazardous substances.

The following chapter is designed to provide a simple introduction to such rules. These should be thoroughly studied, especially by beginners and inexperienced personnel, before starting any laboratory analysis.

Chemical and microbiological studies will become less dangerous if the following basic rules are applied.

2.1.1 Chemical Laboratory

a. Only qualified and experienced personnel should be permitted to work with hazardous reagents and equipment. Beginners must work under the supervision of experienced personnel.
b. A comprehensive safety manual should be read before working in the laboratory.
c. Safety charts must be displayed on laboratory walls to educate personnel with basic safety rules.
d. Chemicals labeled POISON, DANGER, CAUTION, FLAMMABLE, etc. must be handled with extensive care and should not be permitted to come into contact with eyes, mucous membranes, or skin.
e. Mechanical pipettes or a pipette bulb must be used to measure, transfer, and dilute the water/wastewater samples and chemical solutions. Mouth pipetting must not be practiced.
f. The working area must be properly ventilated and equipped with exhaust fans.
g. An efficient fume hood must be employed where hazardous gases such as H_2S, NH_3, or fumes of volatile organic solvents are released.
h. The laboratory should be well equipped with the following safety equipment:

   i. Safety glasses and goggles to protect eyes from chemical splashes or flying objects
   ii. Safety gloves to prevent the exposure of hands and fingers to highly corrosive and toxic materials
   iii. Safety clothing: laboratory coats or aprons should be used to protect skin and clothing from chemical splashes.
   iv. Safety boots to protect feet from spills. Open-toed shoes should not be worn in the chemical laboratory.
   v. Respirators to prevent the inhalation of toxic gases
   vi. Eye wash to clean and wash the eyes in case of an accidental splash

i. All safety equipment should be installed in easily accessible corners or places in the laboratory.

j. Reagent bottles must be replaced on the proper shelves immediately after use.

k. All apparatus required for a specific estimation should be grouped together on the bench to avoid confusion. If possible, the name and procedure written in simple steps should be placed on the wall near the bench.

2.1.2 Microbiological Laboratory

The following additional precautions should be observed while working in microbiological laboratories:

- The microbiological laboratory should be separated from the chemical laboratory. It should be secluded from the passage of other workers and be free from drafts.
- Bench surfaces should be swabbed with disinfectant daily and immediately after accidental spillage.
- Always use mechanical pipettes to prepare inoculum and for subculturing of microbes. Do not use mouth pipetting for this purpose.
- The preparation and transfer of bacteriological samples must be done under a laminar flow cabinet.
- The petri dishes containing colonies of pathogenic bacteria and other hazardous microbes must be sterilized in an autoclave before disposal. These sterilized materials should be collected separately in trash bags and disposed of at sites designated for hazardous waste disposal.
- All cultures and used glassware, including containers and sampling bottles, should be autoclaved before reuse or disposal.
- Boxes and crates for the transport of samples should be made of autoclavable material like metal or plastic, so they can be sterilized or disinfected on a weekly basis.

2.2 Toxic and Hazardous Substances

Some chemicals fall under this category because of their toxic nature and adverse effect on human health. This section describes the possible hazardous substances used in water and wastewater analytical estimations.

While handling these substances, laboratory personnel must follow the safety rules mentioned below:

- Read labels of each reagent bottle or container carefully and follow all the instructions strictly.
- Store these materials in safety cabinets with proper labeling.
- Never remove or cover the label on a reagent bottle or container while it contains reagent.
- Do not use these reagent bottles to store another chemical solution without properly cleaning the bottle and changing its label.
- If the label of a reagent bottle is damaged or hard to read, immediately replace the label and write all the required precautions on it.
- Always handle, store, and transport these materials with extreme care to prevent any contact by inhalation, ingestion, or skin exposure.
- Avoid spillage or splashing into the eyes. If this happens accidentally, follow the First Aid Rules immediately.
- Always prepare standard solutions of these substances; make their dilutions and transfer in a fume hood.
- Do not discharge these substances or their solutions, especially azide-containing compounds, directly into drainpipes. This may cause corrosion or may produce explosive compounds with copper or lead plumbing fixtures.
- Hazardous wastes must be collected and disposed of at sites selected for hazardous waste disposal.

Toxic and hazardous substances can be classified under the following three major categories:

1. Solids and liquids
2. Gaseous and volatile components
3. Biohazards

2.2.1 Solids and Liquids

a. Acids and Alkalis

Concentrated acids (pH $\leq$ 2.0) and alkalis (pH $\geq$ 12.0) are highly corrosive in nature. They may cause burns when they come into contact with skin. They are especially hazardous if spilled or splashed into the eyes.

Cautions: Always store concentrated acids in bottles of 2.5-L capacity, which can be easily lifted and transported.

The dilution of concentrated acid causes a highly exothermic reaction. It is very violent and may cause an explosion if water is added to the acid during dilution. Hence, never add water to the acid. Always add acid slowly and carefully to water.

The dissolution of a base is also an exothermic reaction. Do not hold the beaker or container in hand. While diluting, keep the beaker/container on the stirrer and add water slowly to prepare alkali.

b. Arsenic

Inorganic compounds containing arsenic are used to prepare standard solutions or may be present in the sample itself. Arsenic is highly toxic and may cause lung cancer or death.

c. Azides

Sodium azide is used in different estimations such as detection of dissolved oxygen. It is highly toxic in nature. On acidification, it produces extremely toxic hydrazoic acid.

Cautions: Do not acidify the solutions containing azides.

Destroy azides by adding a sufficient quantity of concentrated solution of sodium nitrite, $NaNO_2$ – 1.5 g $NaNO_2$ per g sodium azide.

d. Cyanides

Cyanides are also used as reagents or may be present in the sample. They are highly toxic in nature. In acidic medium, cyanides produce hydrogen cyanide, an extremely poisonous gas.

Cautions: Do not acidify cyanide solutions.

Add strong alkaline solution of NaOH at the site of a spill because cyanide compounds are stable in strong alkaline solutions.

Always prepare cyanide solutions in a fume hood.

Use self-contained breathing apparatus or supplied air in case of emergency due to spill.

e. Mercury

Mercury and its compounds are used to prepare standard solutions. It serves as an indicator liquid in manometers and thermometers. Mercury is a toxic volatile liquid.

Caution: Upon accidental spillage, immediately cover a mercury spill with enough sulfur powder to minimize volatilization of mercury and then clean.

f. Perchloric Acid

Perchloric acid used as a digesting agent to digest organic matter present in the sample. It can react explosively with organic matter.

Cautions: Samples containing organic matter must be predigested with nitric acid before adding perchloric acid.

Do not add perchloric acid to hot solutions.

Use special perchloric acid fume hood and ducting when digesting the samples with perchloric acid.

2.2.2 Gaseous and Volatile Components

a. Compressed Gases

Compressed gases are widely used in the estimation of metals, organic and inorganic volatile constituents by atomic absorption spectroscopy or gas chromatography. These gases may be flammable or explosive and require careful handling.

Cautions: Protect the cylinders themselves from freezing, overheating, and mechanical damage.

Cylinders must be stored in properly ventilated safety cabinets.

Each cylinder must be supplied with an appropriate pressure gauge and pressure-releasing valves.

Chain, lock, or otherwise prevent the cylinders from moving or falling over.

b. Volatile Organic Compounds

Organic solvents and solid organic reagents are used in various analytical estimations. These are highly volatile compounds, which may be flammable or explosive.

Cautions: Organic solvents, such as benzene and its derivatives like chloroform, carbon tetrachloride, toluene, etc., must be handled in a fume hood.

Store used organic solvent in labeled waste container.

Most waste organic solvents can be repurified by distillation and reused. Distill carefully in distillation apparatus where the temperature does not exceed the boiling point of the solvent concerned.

2.2.3 Biohazards

Contaminated water and wastewater samples contain disease-causing pathogenic bacteria such as Escherichia coli, salmonella, viruses, etc. These must be handled with extreme care. Exposure to these microbes may be incidental during chemical and biological estimation of samples or the examination of microbes producing specific diseases.

Cautions: *Glass pipettes must not be used for microbiological estimation. Always use mechanical pipettes.*

Use aseptic techniques and sterilize all equipment used for sample preparation, inoculation, and transfer of biological samples.

Sterilize the discarded media and cultures before disposal.

Weiss[1] provides more information on toxic and hazardous substances.

2.3 Storage and Transport of Laboratory Chemicals

- Do not store hazardous and corrosive chemicals in breakable containers of more than 2.5-L capacity.
- Do not hold the breakable container only by the neck, but also support from underneath. For transport of containers over longer distances, buckets, baskets, or wooden cases must be used.
- Store all chemicals in cool, dry places. Larger quantities must never be stored in work places or laboratories, but kept in properly ventilated, cool, and dry storage areas.
- Always store flammable gases in pressurized containers and under cold conditions.
- Store organic solvents and liquids in safety cabinets under cold conditions.
- Always store the microbiological media under 4°C in a refrigerator. Do not use them after expiration date mentioned by manufacturer, since they would give false results.
- Reagents labeled POISON, DANGER, CAUTION, FLAMMABLE, etc. must be stored in safety cabinets with proper labeling.

2.4 Fire

2.4.1 Rules for Safety in Case of Fire

- Avoid bumping during distillation and digestion by using antibumping chips or glass beads.
- Never store an easily ignitable liquid in a refrigerator since sparking may cause ignition of vapors.

- Electrostatic charges can start fires through sparking. Charging can occur during pouring of nonconducting liquids like acetone, ether, carbon disulfide, toluene, etc. to glass or plastic containers. Hence, precautions must be taken against free fall of these liquids by using a funnel with a long neck that can reach to the bottom of the container.
- Always take precautions while handling inflammable compounds to prevent spreading of their vapors. Sources of ignition like burners, lamp mantles, and hot plates should not be kept near such substances.
- Take special care during heating of more than 50 ml of inflammable liquid.
- All EXITS must be clearly marked and kept free from any obstacles for workers in the laboratory. All must be aware of locations.
- Each laboratory must be equipped with fire-protective aids like safety showers, fireproof blankets, and fire-protective gels.
- Fire alarms must be installed in each laboratory.
- Emergency call numbers must be clearly marked near each telephone.
- Portable fire extinguishers must be installed at different easily accessible locations in the laboratory.
- All fire protection equipment and materials must be clearly marked in red.
- Training for fire prevention, fire protection, and handling of fire-protection equipment must be given to all technicians.

2.4.2 Rules for Fire Fighting

- Remove the injured person from the fire area.
- If a victim's clothes are burning, rapidly wrap him or her in an extinguishing blanket, spray with CO_2 from a fire extinguisher (not on the face), or place him under the laboratory shower.
- Activate the fire alarm.
- If possible, try to remove the flammable material from the accident site.
- If fire is due to electric short circuit, immediately switch off the main electricity supply.
- Try to extinguish fire, but in uncontrollable conditions immediately evacuate the building and call emergency services and fire brigade.

2.5 First Aid

The laboratory shall be equipped with the following essential items:

- First aid instruction charts
- First aid kits
- Laboratory eye wash kit
- Important telephone numbers of emergency physicians, hospitals, and ambulance

2.5.1 Aid for Burns

In case of excessive burning of any degree, consult a physician.

a. Fire

1. First-degree burns (only redness) — apply oil dressing or a bland ointment or gauze saturated with mineral oil.
2. Second-degree burns (blisters) — cover wound with sterile gauze saturated with olive oil or some burn ointment.
3. Third-degree burns (destruction of tissues) — cover the wound with a sterile dressing and contact a physician immediately.

b. Acids

Wash with large quantities of water, then wash with 5% sodium bicarbonate ($NaHCO_3$) solution and apply dressing.

c. Alkalis

Wash the burnt area with large quantities of water and cover with oil dressing.

d. Hydrofluoric acid

Wash immediately under running cold water and then with 2% ammonia solution or 5% sodium bicarbonate ($NaHCO_3$) solution.

e. Iodine

Wash with 1% sodium thiosulfate solution.

f. Phenol

Wash immediately with alcohol and then apply an oil dressing.

g. Eye burns

Immediately flush eyes with cold water while rolling the eyes in all directions. Contact a physician.

2.5.2 Aid for Cuts

- Do not touch or wash the wound.
- Cover the wound with sterile gauze and paint the surrounding area with 3.5% tincture iodine.

- If the cut is severe and bleeding freely, apply a ligature between the site of injury and the periphery. Cover the wound with sterile gauze until the arrival of a physician.

2.5.3 Aid for Poisoning

a. If the victim is conscious, vomiting should be induced. Activated-charcoal tablets may also be given. Have the victim lie on his or her side until the arrival of a physician.

b. If unconscious, then

   1. Place the victim on his or her side, face toward the floor, and lift head backward.
   2. Monitor the pulse and breathing.
   3. If breathing is interrupted, first blow air strongly into the mouth or nose, then wait for 30 sec. Continue the process of artificial respiration until the arrival of a physician.

Chapter 

Quality Control and Quality Assurance

A good quality control program is essential to obtaining reliable and comparable analytical results. Adequate quality control in analytical estimations depends on the following factors:

- Well trained and experienced personnel
- Good physical facilities and equipment
- Approved quality reagents and standards
- Regular maintenance and calibration of instruments
- A knowledgeable and understanding management

Quality control should begin at the start of sample collection and be carried through the final preparation and documentation of results.

3.1 Sampling

Sampling is the first essential step in assessing the quality of water and wastewater. A representative sample that highlights the exact condition existing in water and wastewater systems must be collected. The data collected from the analysis of the representative sample will serve as a basis for designing the treatment facilities. This will also help to monitor and control water and wastewater treatment processes.

After collection, the sample must be handled and preserved carefully to prevent any alteration in physical, chemical, and biological state. Observations and analysis conducted at the site and in the laboratory provide an adequate basis to assess the quality or extent of contamination or pollution in such a sample. Thus, accuracy and reliability of analytical results depend on the sample collection program. The time, location, type, and frequency of sample collection must be decided according to the source of water and wastewater produced.

Generally, two types of samples are collected — grab and composite.

3.1.1 Grab Sample

A grab sample is one collected at a particular time and place. It represents the condition of the water or wastewater stream at the time of sampling. Grab sampling is preferred under the following conditions:

- The composition of water or wastewater is fairly consistent, i.e., not changing frequently.
- An unusual and intermittent waste is discharged into the water and wastewater collection and treatment facilities. Thus, a grab sample is collected at that particular time to assess the type and source of the unusual discharge.
- When it is necessary to ascertain the quality of wastewater at peak flow. This helps personnel to design the treatment plant to accommodate peak flows. The performance of the wastewater treatment plant could be significantly affected if peak-flow factor is not taken into consideration in designing a treatment plant.
- Industrial wastewater is discharged intermittently to wastewater networks.
- The discharge permit for a treatment plant requires a grab sample.

In addition to the above conditions, a grab sample is required for certain tests that must be performed at the sampling site itself, e.g., chlorine residual, dissolved oxygen, temperature, toxic-gas emission, etc.

3.1.2 Composite Sample

A composite sample is a number of grab samples collected at definite intervals of time over a fixed period (often a 24-hr period) and mixed. This sample presents the average characteristics of water or wastewater flow over that particular period of time. The composite sample is preferred in the following circumstances:

- The average water and wastewater or sludge conditions over a period of time are needed for monitoring and operating a treatment plant.
- Plant efficiencies are to be estimated.
- The generation of malodorous gaseous components in slime layer is to be estimated.

The composite sample may be a fixed-volume or flow-proportioned sample. In the former case, both the time interval and the volume of the sample remain constant, while in the latter case, both time and volume of the sample vary according to the flow of water and wastewater.

For more detailed information on sampling and sampling equipment, see Tchobanoglous and Eliasson.[2]

3.2 Sample Handling and Preservation

Sample handling and preservation are the most important aspects of water and wastewater analysis. Water and wastewater samples are collected for a variety of purposes — to assess the degree of treatment, performance of a treatment facility, extent of pollution, etc. When a sample is collected, during transportation physical, chemical, and biological changes may occur that can alter the sample composition and produce false results. To overcome this, the samples should be handled carefully and preserved if the analysis is to be performed in the laboratory, rather than at the site. For appropriate sample handling, the following precautions should be taken.

3.2.1 Chemical Estimations

- Sampling bottles or containers should be washed with nonphosphate detergent, rinsed thoroughly with running water, and finally rinsed with distilled deionized water (DDW). Traces of chemical or detergent can interfere with the analysis.
- A narrow-mouth sample bottle should be used if any gaseous element is to be analyzed in the sample.
- A wide-mouth sample bottle is used to collect a sample rich in oil or grease content. The wide mouth allows the technician to clean and wipe the interior of the container thoroughly.
- The contents of water and wastewater can continue to alter at different rates after the time of sample collection due to ongoing chemical and biological activities. As only a few parameters can be measured at site during collection, a preliminary treatment or preservation is essential in many cases.

Table 3.1 represents the methods of preservation of samples for the estimation of various parameters. Physical preservation of samples by cooling, i.e., by keeping the sampling container on ice during transportation, is highly essential — especially at high atmospheric temperature. Increase in temperature may cause an increase in the rate of biochemical processes, which will alter the characteristics of the sample.

TABLE 3.1
Standard Preservation Methods for Water and Wastewater Samples

Measurement	Sample Volume (ml)	Container*	Preservative**	Holding Time***
PHYSICAL				
Color	50	P, G	Cool, 4°C	24 h
Odor	200	G only	Cool, 4°C	24 h
Temperature	100	P, G	Determine on site	No holding
Turbidity	100	P, G	Cool, 4°C	48 h
Settleable matter	1000	P, G	None required	24 h

continued

TABLE 3.1 (continued)

Measurement	Sample Volume (ml)	Container*	Preservative**	Holding Time***
Residue				
TSS	100	P, G	Cool, 4°C	7 days
TDS	100	P, G	Cool, 4°C	7 days
VSS	100	P, G	Cool, 4°C	7 days
Total solids	100	P, G	Cool, 4°C	7 days
CHEMICAL - Inorganic				
pH	25	P, G	Determine on site	2 h
Conductivity	100	P, G	Cool, 4°C	28 days
Acidity	100	P	Cool, 4°C	14 days
Hardness	100	P, G	Cool, 4°C	6 months
Alkalinity	100	P, G	Cool, 4°C	14 days
Bromide	100	P, G	Cool, 4°C	24 h
Chloride	50	P, G	None required	28 days
Chlorine residual	200	P, G	Determine on site	No holding
Fluoride	300	P	None required	28 days
Cyanide	500	P, G	Cool, 4°C, NaOH to pH 12	24 h
Iodide	100	P, G	Cool, 4°C	24 h
Sulfate	50	P, G	Cool, 4°C	7 days
Sulfide	500	P, G	2 ml zinc acetate	24 h
Sulfite	50	P, G	Determine on site	No holding
Nitrogen				
Ammonia	400	P, G	Cool, 4°C, H_2SO_4 to pH < 2	24 h
Kjeldahl total	500	P, G	Cool, 4°C, H_2SO_4 to pH < 2	24 h
Nitrate	100	P, G	Cool, 4°C	48 h
Nitrite	50	P, G	Cool, 4°C	24 h
Dissolved oxygen				
Probe	300	G only	Determine on site	No holding
Winkler	300	G only	Fix on site	4–8 h
Ortho – phosphate				
Dissolved	50	P, G	Filter on site, Cool 4°C	24 h
Hydrolyzable	50	P, G	Cool, 4°C, H_2SO_4 to pH < 2	24 h
Total	50	P, G	Cool, 4°C, H_2SO_4 to pH < 2	24 h
Metals				
Dissolved	200	P, G	Filter on site, HNO_3 to pH < 2	6 months
Suspended	200	P, G	Filter on site	6 months
Total	100	P, G	HNO_3 to pH < 2	6 months

TABLE 3.1 (continued)

Measurement	Sample Volume (ml)	Container*	Preservative**	Holding Time***
Mercury Dissolved	100	P, G	Filter on site, HNO_3 to pH < 2	28 days (G), 13 days (hard P)
Total	100	P, G	HNO_3 to pH < 2	28 days (G) 13 days (P)
CHEMICAL - Organic				
BOD	1,000	P, G	Cool, 4°C	24 h
COD	50	P, G	H_2SO_4 to pH < 2	7 days
Oil / grease	1,000	G only	Cool, 4°C, H_2SO_4 or HCl to pH < 2	24 h
Organic carbon	25	P, G	Cool, 4°C, H_2SO_4 or HCl to pH < 2	24 h
Phenol	500	G only	Cool, 4°C, H_3PO_4 to pH < 4, 1.0 g $CuSO_4$	24 h
MABS (Surfactants)	250	P, G	Cool, 4°C	24 h

Adapted from WEF, *Wastewater Sampling for Process and Quality Control MOP OM-1* (1980). Copyright Water Environment Federation. Reprinted with permission.

* P – Plastic and G – glass container. For metals, polyethylene with a polypropylene cap (no liner) is preferred.

** If the sample is stabilized by being cooled to 4°C, it should be warmed to 25°C for experiment, or temperature correction should be made and results reported at 25°C.

*** Holding times listed above are recommended for properly preserved samples.

3.2.2 Microbiological Estimations

For microbiological examinations of water and wastewater samples, the collection, storage, and transport of samples should be carried out under controlled conditions to assess the accurate level of microbial contamination. To achieve this goal, the following precautions are essential:

- Bacteriological sampling bottles must be sterilized in an autoclave at 121°C (15 psi, pressure) for 15 min before each use.
- Every sampling bottle must be properly labeled with information mentioned in Table 3.2.
- Extra care must be taken to avoid any accidental contamination during sampling.
- The sample must not be exposed to light. It must be transported in an insulated container filled with ice.
- The time for sampling, transportation, and estimation should be minimized as far as possible. The estimation should be performed within 6 h of sampling.

- Before inoculation, the sample must attain room temperature.
- If the water or treated wastewater sample contains residual chlorine, an adequate quantity of sodium thiosulfate ($Na_2S_2O_3$) must be added to neutralize it.

TABLE 3.2
Essential Information for Labeling of Sampling Bottles

S. No.	Information	Description
1.	Details of agency requesting analysis	Name Address Contact number
2.	Sampling technician / agency	Name and Address of organization
3.	Sample number and reference	Any alphabetical letter or number
4.	Date of sampling	
5.	Time of sampling	
6.	Place of sampling	Well, river, reservoir, sea, water or wastewater networks, or treatment plants
7.	Purpose of sampling	Routine analysis, assessment of pollution / contamination, public complaint, or disinfection
8.	Type of water sample	Raw or treated effluent, irrigation water, potable, surface, ground, etc.
9.	Residual chlorine (if present)	Concentration in mg/L

3.3 Laboratory Glassware and Reagents

3.3.1 Grade or Quality of Glassware

Material. A high quality or grade of laboratory glassware is essential for all analytical purposes. For general laboratory use, the most suitable material for glassware is borosilicate glass, commercially known as Pyrex®. This material is resistant to all chemicals and can withstand extreme temperatures.

The most common glassware used in water and wastewater testing laboratories are beakers, conical and culture flasks, funnels, cylinders, burettes, pipettes, reagent bottles, test tubes, and culture tubes.

Stoppers. Stoppers, caps, or plugs must be resistant to the effects of materials stored in the container. Different types of stoppers and their recommended uses are as follows:

- Pyrex glass stoppers are commonly used in separatory funnels, BOD bottles, flasks, cylinders, and reagent bottles. The containers used to store strong alkaline solutions should not be closed with glass stoppers because of their tendency to stick, which causes the stopper to freeze into position.

- Cork stoppers wrapped with a relatively inert metal foil are suitable for many samples and reagents.
- Rubber stoppers are best for the containers used to store strong alkaline solutions. These are not recommended for bottles or containers holding organic solvents because they swell and disintegrate when in contact with organic solvent fumes.
- Neoprene stoppers are best suited for containers used to store petroleum products, mineral oils, and other organic materials that degrade natural rubber.
- Teflon (Polytetrafluoroethylene) is the most suitable material to be used for different kinds of reagents, chemicals, and organic solvents.
- Phenolic caps with rubber liners are usually used for culture tubes and flasks because they are resistant to autoclaving.
- Black rubber stoppers are used only for Kjeldahl flasks for nitrogen estimation.
- Metal screw caps must not be used because they are prone to corrosion.

Sampling bottles. Sampling bottles used for the collection of water and wastewater samples should be of good quality glass or plastic and free from toxic substances.

- Bottles used for biological estimation sampling should be fitted with ground-glass stoppers or screw caps fitted with silicone rubber, which will withstand repeated sterilization at 121°C under pressure (15 psi).
- The wide-mouth sampling bottle is used for the collection of samples rich in oil or grease content.
- The narrow-mouth sampling bottle is used for sampling of gaseous components.

Petri dishes. Petri dishes are used for biological examinations and should be of adequate size and depth to ensure constant surface area of the medium for growth and propagation of microbes. These dishes are of two kinds.

1. Glass petri dishes: need sterilization at 121°C under pressure (15 psi) in an autoclave before use.
2. Disposable plastic petri dishes: use directly without sterilization and discard after estimation is done. Use a new one for each estimation.

3.3.2 Reagents

Reagents used for analytical examinations should be of Analytical Reagent (AnalaR) Grade or be known to meet the specifications of purity established by the American Chemical Society (ACS).[3] If a commercial-grade chemical is required in bulk for any analysis, the analyst must make sure that such chemicals do not contain undesirable impurities, which can interfere with the analytical procedure.

3.3.3 Water (DDW)

Water used for blanks, standard solutions, reagents, dilutions, and rinsing of the glassware should be of high purity.

In this manual DDW is employed for all analytical examinations. It is obtained through the following procedure:

1. Tap water is distilled in any commercially available water still. The result is called single distilled water.
2. The single distilled water is then passed through a deionization unit composed of a mixture of Cation–Anion resins. A strong acid resin will remove cations from the water and replace them by H^+ ions. On the other hand, a strong base resin (OH-form) will remove anions. The pH of water so obtained usually lies in the neutral range and conductivity varies from 1 to 2 μ S/cm^{-1}.

3.4 Cleaning of Glassware

3.4.1 Glassware used for Chemical Estimations

The washing of glassware depends on its proposed use. The following list of washing methods is useful for water and wastewater testing laboratory analysts.

Glassware used for routine analysis. This type of glassware should be cleaned and scrubbed with sodium bicarbonate ($NaHCO_3$), rinsed thoroughly, first with tap water and then with DDW.

Glassware used for specific chemical estimations. This type of glassware should be washed with chromic acid, which should be prepared in the laboratory with extreme care according to the following procedure:

- Saturated solution of potassium dichromate ($K_2Cr_2O_7$). Take 100 ml DDW in a beaker or an Erlenmeyer flask of 250-ml capacity. Place on a magnetic stirrer. Add $K_2Cr_2O_7$ in small increments with constant stirring until a saturated solution is prepared.
- Pipette 50-ml saturated solution of $K_2Cr_2O_7$ in a 2-L beaker or an Erlenmeyer flask. Place on the magnetic stirrer.
- Now slowly and carefully add 1.5-L concentrated sulfuric acid to the saturated solution. Stir gently to dissolve.
- Cool the solution. Chromic acid is ready for use.

Cautions: The preparation of chromic acid is a highly exothermic reaction, requiring the following precautions:

- Wear protective clothing, gloves, and safety glasses while making and using this cleaning solution.
- Prepare the solution in a fume hood.

Glassware used for nitrogen and phosphorus estimation. This type of glassware should be washed in the following manner:

- Wash with $NaHCO_3$ and rinse three to four times with tap water and finally with DDW.
- Soak overnight with 6 N HCl. Then wash thoroughly and rinse with DDW.
- Do not use detergent for this purpose.

3.4.2 Glassware Used for Microbiological Estimation

The following precautions should be taken for cleaning of this type of glassware:

- Sampling bottles, culture tubes and flasks, and glass petri dishes should always be cleaned with Na_2CO_3, rinsed in running tap water, and then in DDW. Finally, the cleaned and washed glassware should be sterilized in an autoclave at 121°C and 15 psi pressure for 15 min. If this glassware is used with ground-glass stoppers, a thin strip of paper or foil must be inserted between the stopper and the neck of the glassware before placing in the autoclave. This will prevent jamming of the stopper or cracking of the glass on cooling.
- Pipettes used for biological transfers should be placed in a jar of disinfectant having free chlorine such as hypochlorite, after every use. They must be thoroughly washed with fresh water and finally rinsed with DDW. Dry and pack them in a metal canister and sterilize in an autoclave. After sterilization, plug the pipettes with cotton wool to prevent contamination.
- Culture tubes and flasks should be plugged with cotton wool or closed with polypropylene caps. The necks and stoppers of bottles should be covered with paper or foil to prevent contamination.

3.4.3 Spectrophotometric Cells

The cleaning of spectrophotometric cells needs the following precautions:

- Clean the cells with a mild agent or sulfonic detergent that does not contain particulate matter.
- The cleaning of cells must be done as soon as possible after each use.
- Always use DDW for rinsing if aqueous solutions are measured in the cell.
- Use a mixture of equal parts 3N HCl and ethanol to remove any stains or deposits of oily or greasy substances.
- If any organic material is measured in the cell, use a suitable organic solvent for cleaning.
- Use soft tissue to wipe the cells dry. Never blow air to dry the cell.
- Never clean the cells with a brush or any other cleaning device, which may scratch the walls of the cells.
- Never use hot concentrated acid or alkali for cleaning.
- Never clean the cells with ultrasonic devices.

3.5 Preparation of Standard Solutions

3.5.1 Important Terms for Expressing the Concentration of a Solution

Equivalent weight. Equivalent weight of a substance is that weight that combines with or displaces one part by weight of hydrogen or 8 parts by weight of oxygen or 35.5 parts by weight of chlorine in a chemical reaction.

Equivalent weight of an acid. That weight of an acid that contains 1 part by weight of replaceable hydrogen. This can be expressed as:

$$\text{Equivalent weight of an acid} = \frac{\text{Gram molecular mass}}{\text{Basicity of acid}}$$

Basicity of an acid. The number of H^+ ions furnished by each molecule of an acid in aqueous solution.

For example, basicity of HCl = 1, basicity of H_2SO_4 = 2, and basicity of acetic acid (CH_3COOH) = 1.

$$\text{Equivalent weight of HCl} = \frac{36.5}{1.0} = 36.5$$

$$H_2SO_4 = \frac{98}{2.0} = 49.0$$

Equivalent weight of a base. That weight of a base that reacts with 1 part by weight of hydrogen.

$$\text{Equivalent weight of a base} = \frac{\text{Gram molecular mass}}{\text{Acidity of base}}$$

Acidity of base. The number of OH^- ions furnished by each molecule of base that react with H^+ ions.

For example: Acidity of NaOH = 1 and Acidity of $Ca(OH)_2$ = 2

$$\text{Equivalent weight of NaOH} = \frac{40}{1.0} = 40 \text{ and } Ca(OH)_2 = \frac{74}{2} = 37$$

Equivalent weight of oxidizing or reducing agents. The equivalent weight of an oxidizing or reducing agent can be calculated in either of the following ways:

a. Number of electrons gained or lost during a Redox reaction.

$$\text{Equivalent weight of oxidizing/reducing agents} = \frac{\text{Molecular Weight}}{\text{No. of electrons}}$$

i. Example:

$$KMnO_4 \text{ in dilute } H_2SO_4 — MnO_4^- + 8\ H^+ + 5\ e \leftrightarrow Mn^{2+} + 4\ H_2O \quad (3.1)$$

In this reaction $KMnO_4$ utilizes 5 electrons. Hence

$$\text{Equivalent weight of } KMnO_4 = \frac{\text{Molecular weight of } KMnO_4}{5}$$

ii. Example:

$$K_2Cr_2O_2 \text{ in dilute } H_2SO_4 — Cr_2O_7^{2-} + 14\ H^+ + 6\ e \leftrightarrow 2\ Cr^{3+} + 7\ H_2O \quad (3.2)$$

In this reaction, $K_2Cr_2O_7$ utilizes 6 electrons. Hence,

$$\text{Equivalent weight of } K_2Cr_2O_7 = \frac{\text{Molecular weight of } K_2Cr_2O_7}{6}$$

b. Equivalent weight of oxidizing or reducing agent can be calculated by the change in its electronic charge during a chemical reaction. This electronic charge represents its oxidation state, which is known as oxidation number and denoted as O.N.

Oxidation number of an element in a compound is defined as a number indicating the amount of oxidation or reduction required to convert that element from its free state to that state in which it is present in the compound. If the O.N. change is Z, the equivalent weight can be calculated as follows:

$$\text{Equivalent weight of oxidizing or reducing agent} = \frac{\text{Molecular weight}}{Z}$$

Example:

i. $KMnO_4$ in dilute H_2SO_4 –

Mn element in $KMnO_4$ in the presence of dilute H_2SO_4 is reduced to Mn (II) salt.

$$\overset{+1}{K}\ \overset{+7}{Mn}\ \overset{-8}{O_4} \rightarrow \overset{+2}{Mn}\ \overset{+6}{S}\ \overset{-8}{O_4}$$

The O.N. of Mn changes from +7 to +2 in the above reaction; therefore, Z = –5.

$$\therefore \text{Equivalent weight of } KMnO_4 = \frac{\text{Molecular weight of } KMnO_4}{5}$$

Similarly, in Equation, 3.2 the O.N. of Cr changes from +12 to +6, therefore Z= –6.

$$\therefore \text{Equivalent weight of } K_2Cr_2O_7 = \frac{\text{Molecular weight of } K_2Cr_2O_7}{6}$$

The equivalent weights of some oxidizing or reducing agents commonly used in volumetric determinations are given in Table 3.3.

TABLE 3.3
Equivalent Weight of Some Common Oxidizing and Reducing Agents

Agent	Nature	Condition of Reaction	Oxidation State Change (Z)	Equivalent Weight
Potassium permanganate, $KMnO_4$	oxidizing	Acid	5	Mw / 5
Potassium permanganate, $KMnO_4$	oxidizing	Alkaline	3	Mw / 3
Potassium dichromate, $K_2Cr_2O_7$	oxidizing	Acid	2 x 3	Mw / 6
Iodine, I	oxidizing	Acid	1	Mw / 1
Potassium iodide, KI	reducing	Acid	1	Mw / 1
Sodium thiosulfate, $Na_2S_2O_3$	reducing	Acid	1*	Mw / 1*
Ferrous ammonium sulfate, $Fe(NH_4)_2(SO_4)_2$	reducing	Acid	1	Mw / 1

Adapted from Sawyer, C. N. and McCarty, P. L., *Chemistry for Environmental Engineering*, 3rd ed., McGraw-Hill Inc., 1978, with permission.

* The equivalent weight of $Na_2S_2O_3$ must be calculated indirectly because the oxidation number change is not certain. In the reaction with iodine, one $Na_2S_2O_3$ is equivalent to one atom of iodine; therefore, the equivalent weight is MW/1.

Normal solution. The solution that contains one gram equivalent weight of a substance per liter of solution is known as normal solution.

Example: 1N solution of NaOH —

Gm. molecular weight of NaOH = 40 g

Acidity of NaOH = 1

$$\therefore \text{Equivalent weight of NaOH} = 40$$

When 40 g of NaOH is dissolved in DDW and the final volume of solution is made up to 1L, this solution is known as 1N solution of NaOH.

Normality. Normality of a solution is defined as the number of gram equivalents of a substance dissolved per liter of the solution and is represented by the symbol N.

$$\text{Normality, N} = \frac{\text{Gram equivalent of substance}}{\text{Volume of solution, L}}$$

Molar solution. The solution that contains gram molecular weight of a substance dissolved in 1 L of solution is called molar solution.

Example: 1M solution of glucose—

$$\text{Gm. molecular weight of glucose, } C_6H_{12}O_6 = 180 \text{ g}$$

When 180 g glucose is dissolved in DDW and final volume of solution is made up to 1L, this solution is known as 1M solution of glucose.

Molarity. Molarity of a solution is defined as the number of gram moles of the substance dissolved per liter of the solution and is represented by the symbol M.

$$\text{Molarity, M} = \frac{\text{Gram moles of substance}}{\text{volume of solution, L}}$$

Relation between normality and molarity of a solution.

$$\text{Normality} = \text{molarity} \times \frac{\text{Molecular mass of substance}}{\text{Equivalent mass of substance}}$$

For acids: normality = molarity × basicity of acid

For bases: normality = molarity × acidity of base

Mass percentage (w/v). Mass percentage is defined as the number of parts by mass of substance per hundred parts by volume of solution.

Example: 1 g NaOH is dissolved in DDW. The final volume of solution is brought up to 100 ml. This solution is said to be 1–% (w/v) NaOH solution.

Volume percentage (v/v). Volume percentage is defined as the number of parts by volume of a substance per hundred parts by volume of solution.

Example: If 1 ml HCl is mixed with DDW and final volume is brought up to 100 ml, the solution is said to be 1% (v/v) HCl solution.

For more detailed information, see Jeffery et al.[4]

3.5.2 Preparation of Solutions

3.5.2.1. Stock Standards

To prepare the stock solution of reagents, the following standard procedure should be followed:

- Take a sufficient quantity of anhydrous chemical in a clean, dry beaker and place it in an oven for a few hours before use. Usually the oven is set at 103°C, if temperature is not specified. After drying, cool the chemical in a desiccator until it attains room temperature.
- Use an analytical balance with readability 0.01-mg to weigh the chemical accurately. Set the balance at 0 and place a weighing paper on the pan. Adjust the balance again at 0. Place the chemical slowly on the weighing paper and measure accurately till a constant weight is achieved.
- Transfer the accurately weighed chemical in a prerinsed volumetric flask with the help of a clean and dry funnel. Add some fresh DDW. Also rinse the paper with DDW. Add more DDW so that the flask is about three-quarters full. Carefully swirl or stir the solution until the chemical is dissolved. Fill the volumetric flask so that the meniscus (flat surface) of the solution is at the calibration point. Close the flask with suitable stopper. Invert the flask several times to mix the solution thoroughly.
- Store the stock standard solutions in properly cleaned and dry containers under refrigeration. Follow any specific instructions given for the test such as using a dark container or storage away from light.

3.5.2.2 Solution of Particular Normality or Molarity

- Before starting the preparation, always keep the stock solution outside the refrigerator until it attains room temperature.
- Take a clean, dry volumetric pipette and rinse with the solution three–four times before pipetting the specified amount of the stock solution.
- Use volumetric flasks for dilution of stock solutions to prepare a solution of particular normality or molarity. For all analyses, rinse the volumetric flask with DDW prior to pipetting the stock solution. For nitrogen and phosphorus analysis, rinse the volumetric flask first with 6N HCl and then with DDW.
- Bring the final volume of solution up so that its meniscus lies at the calibration mark.
- Now close the volumetric flask with a glass stopper and twist the stopper slightly to achieve a secure fit.
- Mix the solution thoroughly by inverting the flask several times.

3.5.3 Preparation of Acids and Alkalis

3.5.3.1 Acid Solutions

- Measure 500 ml DDW in a 1L volumetric flask.
- Add the required amount of concentrated acid (see Table 3.4) to DDW with extreme caution.
- Fill DDW up to the mark to bring the final volume to 1L.
- Mix the solution well and cool.

TABLE 3.4
Preparation of Acid Solutions of Different Normality

Characteristics	Hydrochloric acid (HCl)	Sulfuric acid (H_2SO_4)	Nitric acid (HNO_3)
Specific gravity (20/4°C) ACS grade concentrated acid	1.174–1.189	1.834–1.836	1.409–1.418
Percent of active ingredient in concentrated acid	36–37	96–98	69–70
Normality of concentrated acid	11–12	36	15–16
Volume (ml) of concentrated acid to prepare 1 L of:			
18 N solution	-	500	-
6 N solution	500	167	380
1 N solution	83	28	64
0.1 N solution	8.3	2.8	6.4

Adapted from American Public Health Association, *Standard Methods for the Examination of Water and Wastewater,* APHA, 17th ed., 1989, with permission.

Note: *Always add concentrated acid to DDW. The reverse procedure will be violent and could result in an accident. Use protective gloves while making the solution and prepare it in a fume hood.*

3.5.3.2 Alkaline Solutions

- To prepare the standard solutions of NaOH or KOH, boil about 1L of DDW for a few minutes (to expel CO_2 gas) in a beaker. Cover the beaker and cool to room temperature.
- Weigh the required amount of NaOH or KOH (see Table 3.5) and transfer to a volumetric flask. Add about 600 ml boiled and cooled DDW in the flask. When the solute is dissolved, allow the solution to cool down. Make up the final volume to 1L with remaining DDW.

Note: *(1) Do not hold the flask in hand, as the preparation of alkaline solution is an exothermic reaction.*

TABLE 3.5
Preparation of Alkaline Solutions of Different Normality

Normality of Hydroxide Solution*	Weight of Sodium Hydroxide (NaOH, g/L)	Weight of Potassium Hydroxide (KOH, g/L)
6 N	240	336
2 N	80	112
1 N	40	56
0.1 N	4	5.6

Adapted from American Public Health Association, *Standard Methods for the Examination of Water and Wastewater*, APHA, 17th ed., 1989, with permission.
* The normality of these alkaline solutions must be checked before use by acid – alkali titration.

(2) Store the stock alkaline solutions in polyethylene bottles with polyethylene screw caps. Never store strong alkaline solutions (NaOH or KOH) in glass bottles with glass screw caps, as the glass will become etched and the glass stopper frozen in its position.

(3) The solutions of NaOH and KOH must always be standardized before use with acid–alkali titration.

3.5.4 Preparation of Acids and Alkalis of Particular Normality from Stock Solution

When a stock solution of acid/alkali with known normality is available, a solution of particular normality can be prepared by using the relationship given below:

$$N_1 \times V_1 = N_2 \times V_2$$

Where N_1 = Normality of stock solution
V_1 = Volume of stock solution
N_2 = Normality of particular solution
V_2 = Volume of particular solution

Example: If it is desired to make 100 ml of N/5 acid solution from a stock solution of 1N acid strength, the calculation will be as follows:

$$1N \times V_1 = {}^1/_5 N \times 100$$

$$V_1 = 20 \text{ ml}$$

Twenty ml of 1N acid should be diluted to 100 ml with DDW and thoroughly mixed to obtain N/5 strength of acid. This solution should be standardized against a weighed sample of a primary standard.

3.6 Expression of Results

In water and wastewater analysis, the physical and chemical constituents of the sample can be expressed in the following units.

3.6.1 Milligrams per liter: mg/L

Milligrams per liter (mg/L) is a weight–volume relationship, i.e., the weight of physical or chemical constituent in unit volume of water or wastewater sample. This offers a convenient basis for calculations of suspended and dissolved matter in the sample. Furthermore, mg/L is directly applicable to the metric system.

$$\text{mg/L} = \text{g/m}^3 \quad \text{and} \quad \text{g/L} \times 10^3 = \text{Kg/m}^3 \tag{3.3}$$

Sometimes the concentration of pollutant is expressed in lb./million gallon. It is related to mg/L by the following expression:

$$\text{mg/L} \times 8.34 = \text{lb./million gallon} \tag{3.4}$$

3.6.2 Parts per million: ppm

Parts per million is a weight to weight ratio. This is mainly used to measure the gaseous pollutants in air and water.

3.6.3 Relation between mg/L and ppm

The units mg/L and ppm are related by the following:

$$\text{ppm} = \frac{\text{mg/L}}{\text{specific gravity of sample}} \tag{3.5}$$

In the case of water, specific gravity is 1, therefore

$$\text{ppm} = \text{mg/L (in water)}$$

The wastewater collected from domestic resources has specific gravity 1, therefore

$$\text{ppm} = \text{mg/L (in domestic wastewater)}$$

This relation is not true for industrial wastewater because it differs from domestic wastewater in specific gravity depending on the process and the material produced.

So the relation shown in expression 3.5 is used to convert the units in the case of industrial effluent.

3.7 Accuracy and Precision

Accuracy refers to the correctness of the measurement, i.e., how closely the concentration of a pollutant is measured by the test procedure to a known concentration present in the sample.

Precision represents the reproducibility of the method used for measurement. Reproducibility means the sets of experiment are repeated several times on different days and the results are compared to assess the accuracy of the test procedure. To achieve a high degree of accuracy and precision in analytical estimations, the error of experiment should be minimized.

3.7.1 Minimization of Errors[4]

The following methods can minimize the errors in analytical estimations:

3.7.1.1 Calibration of Apparatus

- All meters, such as pH, conductivity, oxygen, etc., must be calibrated on a regular basis according to the instructions of the manufacturer.
- The weight of flasks, beaker, crucible, or other containers should be constant and stabilized before reporting them.
- The calibration of volume measurement with burette and pipette must be done before the data analysis.

3.7.1.2 Running a Blank

This consists of carrying out the determination of a blank prepared with the same volume of DDW instead of the sample. The blank is run along with the sample under exactly the same estimation conditions.

3.7.1.3 Running a Sample of Known Concentration

The accuracy of results can be examined by preparing a synthetic sample with a known amount of pure chemical and analyzing it with a blank under exactly the same estimation conditions.

3.7.1.4 Running Samples in Replicate

The random error of estimation can be reduced considerably by running the sample in replicate, taking a large number of measurements, and calculating the standard deviation of the results.

3.7.2 Standard Deviation (σ)

Standard deviation is calculated from the several individual results of a sample concentration taken at different times under similar experimental conditions represented as $x_1, x_2, x_3 \ldots x_n$, the number of readings taken n times, and mean of the results ($\bar{x}$). These terms are related as follows:

$$\bar{x} = (x_1 + x_2 + x_3 + \ldots x_n)/n = \Sigma x_n/n \quad (3.6)$$

If the deviation of each result of a sample from the mean is represented by d, then the standard deviation (σ) can be calculated as follows:

$$d_1 = x_1 - \bar{x} \quad d_2 = x_2 - \bar{x}$$

$$d_3 = x_3 - \bar{x} \quad d_n = x_n - \bar{x}$$

Then

$$\sigma^2 = (d_1^2 + d_2^2 + d_3^2 + \ldots d_n^3)/n \quad (3.7)$$

In Equation 3.7, σ^2 is known as variance. The square root of the variance gives the standard deviation denoted by σ.

$$\sigma = \sqrt{(d_1^2 + d_2^2 + d_3^2 + \ldots d_n^3)/n} \quad (3.8)$$

It is tedious to calculate the deviation for an individual sample from the mean value $\bar{x}$ and σ^2. The standard deviation can be readily calculated from the sum of the values of x, i.e., (Σ x) and the sum of the values of x^2 that means Σx^2 using the following expression:

$$\sigma^2 = \Sigma x^2 - (\Sigma x)^2/n \div (n - 1) \quad (3.9)$$

$$\sigma = \sqrt{\Sigma x^2 - (\Sigma x)^2/n \div (n - 1)} \quad (3.10)$$

Relation 3.11 can calculate the Standard Error of Mean (SEM):

$$SEM = \sigma/\sqrt{n} = \sqrt{2\ S^2/n} \quad (3.11)$$

where

$$S^2 = \Sigma x^2 - (\Sigma x)^2/n$$

Table 3.6 presents the method to calculate standard deviation (σ) and standard error of mean (SEM) for the analysis of SO_4^{2-} anion in wastewater. For more details on statistical analysis, consult Snedecor and Cochran.[5]

TABLE 3.6
Replicate Analysis of Sulfate in Wastewater

Replicate Samples	Sample 1 (SO_4), mg/L	Sample 2 (SO_4), mg/L	Total
1	302	233	
2	298	235	
3	298	230	
4	301	232	
Σx	1199	930	2129
$\bar{x}$	299.8	232.5	
Σx^2	359,413	216,238	
$(\Sigma x)^2/n$	359,400	216,225	
$S^2 = \Sigma x^2 - (\Sigma x)^2/n$	13	13	26
$n - 1$	3	3	6
σ	$13 \div 3 = \pm 4.33$	$13 \div 3 = \pm 4.33$	

Standard Error of the Mean:

$$\text{Pooled } S^2 = 26 \div 6 = 4.33\text{ ; SEM} = \sigma = \sqrt{2\,S^2/n} = \sqrt{2 \times 4.33/4} = \pm 1.47$$

Chapter 4

Types of Water — Sources and Quality Assessment

The quality of water depends mainly on its source of origin. It varies if and when its source is changed. The quality of water is assessed on the basis of laboratory analysis of various parameters. Thus, the terms "source" and "laboratory analysis" are interdependent. Hence, for a water analyst, the most important analytical aspect is the source and location of water sampling. Based on this fact, the types of water existing in the environment can be broadly classified under following categories:

1. Surface water: collected as runoff from rainwater in ponds, rivers, lakes, etc. Another source is the sea. Sea water is referred to as marine water.
2. Ground water: collected underground by seepage of rainwater and surface water through soil. It is also known as brackish water.
3. Potable water: water suitable for human consumption.
4. Wastewater: a byproduct of various domestic and industrial activities.

4.1 Surface Water

4.1.1 Definition and Uses

The water from rainfall collected in lakes, ponds, and rivers etc. is called surface water. It may be either flowing or still. The public has various uses for surface water, such as.

- Drinking
- Bathing and recreational activities like swimming, skiing, boating, etc.
- Commercial fish farming
- Recreational fishing
- Irrigation

4.1.2 Quality Assessment

Surface water has the unique quality of self-purification, which means the exchange of gases (O_2–CO_2 system) in the environment. This property causes the oxidation of natural or foreign organic pollutants by oxygen dissolved in waters. In oxygen-rich zones, degradable organic substances are broken down into simpler unobjectionable products by the action of aerobic microorganisms.

The discharge of domestic and municipal wastewater, industrial and agricultural wastes (organic, inorganic pollutants, and heat), solid and semi-solid refuse to the stream results in the accumulation of sediments in waterbeds. This reduces the rate of self-purification due to depletion of dissolved oxygen level in water, which in turn initiates the anaerobic decomposition of sediments. The development of anaerobic conditions at waterbeds produces different volatile malodorous components that cause odor nuisance and health hazards. Therefore, precautions should be taken to restrict the discharge of such wastes into streams, rivers and lakes. Otherwise, surface waters must not be used for public consumption. The quality of municipal, industrial and agricultural wastes should be controlled at the point of production. This means that they should be treated before being discharged into surface waters.

Physical, chemical and microbiological analytical estimations provide important information on the quality of water. This data also represents the actual state of pollution of surface water. The estimation of indicator organisms including pathogenic bacteria presents the extent of contamination in surface waters. The common indicator and pathogenic bacteria in water are:

1. Indicator Organisms: Total and fecal coliforms
   Fecal streptococci
2. Pathogenic Organisms: Salmonella
   Shigella
   Intestinal virus
   Parasites

In assessing the quality of surface water, it should be considered that still or stationary water is more susceptible to pollutants than flowing water because of eutrophication (the natural aging process). Eutrophication makes the water rich in nutrients such as phosphorus and nitrogen, which aggravate the growth of algal bloom. Excessive growth of algae takes place when the concentration of P and N exceeds 0.3 mg/l and 0.01 mg/l respectively.[6] Die-off and settling plant growth result in a sediment oxygen demand that tends to reduce dissolved oxygen level. The effect of eutrophication becomes more complicated by large day and night variances in dissolved oxygen due to photosynthesis and respiration. For monitoring the quality of surface water and to assess the extent of contamination, the parameters to be measured are given in Table 4.1.

TABLE 4.1
Essential Parameters for the Analysis of Surface Water

Parameters	General Analysis	Used for bathing and recreation	Analysis for contamination and toxicity
Physical Characteristics			
Odor & color	X	X	X
Turbidity	X	X	X
Temperature	X	X	X
Suspended solids	X	X	X
Dissolved solids	X		X
Residue and floatable substances		X	X
Chemical Characteristics			
pH value	X	X	X
Conductivity	X	X	X
Dissolved oxygen	X	X	X
Total organic carbon (TOC)	X	X	X
Hardness	X		
Carbon dioxide	X		
Trace Elements			
Arsenic	X	X	X
Boron	X		X
Cadmium			X
Lead	X	X	X
Mercury	X	X	X
Nickel		X	X
Zinc			X
Organic Traces			
Anionic surfactants	X	X	X
Mineral oil		X	X
Phenol content	X	X	X
Hydrocarbons	X	X	X
Pesticides		X	X
Inorganic Constituents			
Ammonia	X	X	X
Calcium	X		
Chloride	X	X	X
COD (chemical oxygen demand)	X	X	X
Fluoride	X		X
Iron	X		
Magnesium	X		
Manganese	X		X
Nitrate	X		X
Sodium	X	X	
Sulfate	X		X
Sulfide	X	X	X
Total nitrogen	X	X	X
Potassium	X	X	X

continued

TABLE 4.1 (continued)

Parameters	General Analysis	Used for bathing and recreation	Analysis for contamination and toxicity
Biological Parameters			
BOD (biological oxygen demand)	X	X	X
Total count	X	X	
Fecal coliform & F. streptococci	X	X	X
Salmonella	X	X	X
Intestinal virus	X	X	X
Multicellular organism		X	X

4.2 Ground Water

4.2.1 Definition and Uses

The water that falls on the earth in the form of rain, percolates through the soil, and occupies subterranean permeable layers, is known as ground water. This water occupies the spaces between the soil particles of an aquifer, a water-bearing stratum or formation. The rate of percolation depends mainly on the porosity of soil. Generally, the rate of flow in sand is 1 to 5 m per day, in gravel 6 to 10 m per day, and in silt and clay, the rate can be as low as a few mm or cm per day.

Ground water is utilized primarily as raw water for drinking, as cooling water in industries, and in irrigation.

The great advantage of ground water is its use for drinking purposes without any treatment. Hence, considerable care should be taken to protect ground-water aquifers from irreparable damage through the disposal of waste materials.

4.2.2 Quality Assessment

The quality of ground water is influenced mainly by the quality of its source. Changes or degradation in the quality of source waters can seriously affect the quality of ground-water supply. Municipal- and industrial-waste seepage into an aquifer is a major source of both organic and inorganic pollution.

The introduction of organic pollutants in significant quantities in ground water is very unlikely because of their macromolecular structure, which restricts their movement through the soil. The water may absorb gases of decomposition of biodegradable organic matter, such as H_2S and methane, within the pores of the soil mantle through which it is percolating.

On the other hand, inorganic pollutants can get easy access through the soil, and, once introduced, they are very difficult to remove. The effect of such contamination may continue for a longer period since natural dilution is a slow process and artificial removal or treatment of ground water is impractical or very expensive. Thus, ground water possesses a high level of dissolved salts.

In addition to this, municipal and industrial wastes add biological pollutants, mainly pathogenic microbes, to ground water. But the number of these organisms can be reduced to tolerable levels by the percolation of water through the different layers of soil.

The analysis of ground water differs according to its use. For example:

- The content of pathogenic microbes such as E. coli, fecal coliform, and viruses must be determined when untreated water is used for drinking.
- Salt content and conductivity must be analyzed when ground water is used as cooling water in industries.
- Hardness, nitrate, phosphorus content, SAR, pathogenic bacteria must be examined when ground water is used for irrigation.

The analysis chart given in Table 4.2 represents the essential parameters for analysis to ascertain the quality of ground water when utilized for different activities.

TABLE 4.2
Essential Parameters for the Analysis of Ground Water Utilized in Various Activities

Parameters	General Analysis	Analysis for concrete corrosion	Analysis for contamination & toxicity	Analysis when used for Public supply
Physical Characteristics				
Odor, color, turbidity	X	X	X	X
Temperature	X		X	X
Solids	X		X	X
Residue				X
Ash				X
Chemical Characteristics				
pH value	X	X	X	X
Conductivity	X		X	X
Dissolved oxygen	X	X	X	X
Total organic carbon (TOC)			X	X
Hardness	X	X		X
Trace Elements				
Arsenic			X	X
Boron			X	X
Cadmium			X	X
Copper			X	X
Lead			X	X
Mercury			X	X
Nickel			X	X
Zinc			X	X
Organic Traces				
Hydrocarbons			X	X

continued

TABLE 4.2 (continued)

Parameters	General Analysis	Analysis for concrete corrosion	Analysis for contamination & toxicity	Analysis when used for Public supply
Halogenated hydrocarbons			X	X
Polycyclic hydrocarbons			X	X
Inorganic constituents				
Ammonium	X	X	X	X
Calcium	X	X		X
Chloride	X	X	X	X
Cyanide			X	X
Fluoride	X		X	X
Iron	X	X		X
Magnesium	X	X		X
Manganese	X		X	X
Metaboric acid	X			X
Nitrate		X	X	X
Nitrite				X
Phosphate	X		X	X
Potassium	X			X
Silicic acid				X
Sodium	X			X
Sulfate	X	X	X	X
Sulfide		X	X	X
Biological parameters				
Total count				X
Fecal indicators			X	X
Multicellular organism			X	X

4.3 Potable Water

4.3.1 Definition and Uses

Potable water is drinking water and the most important material for human consumption. Hence, it must be free from any sort of contamination, undesirable substances, and pathogenic bacteria hazardous to health and environment. Raw water for drinking can be obtained from the following sources:

- Ground water and ground water enriched by surface water
- Filtered surface water
- Ground water enriched by purified wastewater
- River water and desalinated sea water
- Natural lakes and reservoirs
- Rain water and spring water

4.3.2 Quality Assessment

A variety of pollutants and contaminants that may conceivably find their way into public water supplies are:

- Toxic substances leached from mineral formation such as fluorapatites
- Phytoloxins produced by some algae growing in surface water
- Heavy metals dissolved from waterworks structures, mainly metallic pipes and water-treatment chemicals like some polyelectrolytes
- Toxic compounds and pathogenic microbes from industrial, domestic and hospital wastes discharged into a watercourse without treatment
- Radioactive substances from fall-out and from the nuclear-energy industry
- Pesticides and insecticides used for pest control in agricultural activities discharged to water bodies along with agricultural wastes

Thus different problems can arise during processing and distribution of potable water. In addition, environmental contamination and discharge of municipal sewage to the drinking-water supply can spread epidemics such as typhoid, cholera, hepatitis, etc. For this reason, regular biological examination of potable water is essential.

The scope of analytical examination must take into account the type of raw water used for drinking purposes. The minimum parameters to be tested to assess the quality of drinking water are given in Table 4.3.

TABLE 4.3
Essential Parameters to Assess the Quality of Water used for Drinking

Parameters	Drinking	Palatability	Corrosivity	Water treatment	Economic significance
1. Physical					
Color	X	X			X
Odor & taste	X	X	X	X	X
Temperature	X	X	X	X	X
Turbidity	X	X	X	X	X
Total suspended solids	X	X			
Dissolved solids	X	X	X	X	
Residue	X		X	X	
2. Chemical					
pH	X	X	X	X	X
Conductivity	X		X	X	X
Alkalinity	X	X	X	X	X
Hardness	X		X	X	X
Dissolved oxygen	X		X	X	X
Carbon dioxide	X		X	X	X

continued

TABLE 4.3 (continued)

Parameters	Drinking	Palatability	Corrosivity	Water treatment	Economic significance
3. Inorganics					
Chloride	X				
Iodide	X				
Fluoride	X				
Ammonia	X			X	
Nitrite	X				
Nitrate	X				
Total nitrogen	X				
Phosphate	X		X	X	
Residual chlorine	X	X		X	
Sulfate	X				
Sulfite	X	X	X		
Sulfide	X	X	X	X	X
Silicates	X		X	X	
COD	X				
Aluminum (Al)	X	X	X		
Barium (Ba)	X				
Sodium (Na)	X	X		X	
Calcium (Ca)	X	X			X
Magnesium (Mg)	X	X			X
Potassium (K)	X	X		X	
Manganese (Mn)	X	X			
Iron (Fe)	X	X	X	X	X
Copper (Cu)	X	X	X		X
Zinc (Zn)	X				
4. Organics					
Chlorinated - hydrocarbon	X				
Phenols	X	X			
Choloropenoxy	X				
Radioactive elements	X				
Surfactants	X				X
Grease/oil	X	X			X
5. Trace Elements					
Arsenic (As)	X	X			
Boron (B)	X	X			
Cadmium (Cd)	X	X			
Chromium (Cr)	X	X			
Cyanide (CN)	X	X			
Lead (Pb)	X	X			
Mercury (Hg)	X	X			
Selenium (Se)	X	X			
6. Microbiology					
BOD	X	X	X	X	
Total count	X	X		X	
E. coli	X	X		X	
Total coliforms	X	X		X	
Pathogens	X	X		X	
Parasites	X	X		X	
Virus	X	X		X	

4.4 Wastewater — Domestic and Industrial

4.4.1 Definition

Water produced by different domestic and industrial activities is known as wastewater. It contains various inorganic, organic and biological contaminants that are of environmental significance. It can significantly pollute ground and surface waters when discharged without treatment. The greatest contamination is caused by the drainage systems from towns, industrial sites, and agriculture. These contaminants can create odor and health hazards if discharged without proper care and treatment into streams or oceans.

TABLE 4.4
Major Wastewater Pollutants and their Impact on Environment

Pollutants	Source	Impact
1. Physical		
a. Suspended solids	Public water supply, domestic & industrial wastes, soil erosion, infiltration / inflow	Can cause anaerobiosis due to sludge deposits at water beds and result in the generation of malodorous components
b. Dissolved solids	Public water supply, domestic & industrial wastes, soil erosion, infiltration / inflow	Impart hardness to water and restrict reuse of treated effluent for irrigation purposes
2. Chemical - Inorganic		
a. Nutrients (N & P)	Domestic, industrial, and agricultural wastes, natural runoff	May cause eutrophication resulting in the excessive growth of algae. When discharged in higher concentrations on land they may lead to the pollution of ground water also.
b. Trace metals	Industrial wastes from mining, petroleum, metal foundries	Mostly toxic in nature thus disturbs the ecological balance in treatment process of wastewater. Restrict the reuse of treated effluent.
c. Gaseous inorganics	Domestic and industrial wastes	Hydrogen sulfide and ammonia create odor nuisance and health hazards
3. Chemical - Organic		
a. Refractory /trace organics e.g. phenols, surfactants etc.	Industrial and agricultural wastes	Resistant to biodegradation hence resists conventional methods of wastewater treatment. May cause taste and odor nuisance, may be toxic or carcinogenic
b. Biodegradable organics such as carbohydrates, proteins and fats	Domestic and industrial wastes	Cause biological degradation at the expense of dissolved oxygen, which may deplete dissolved oxygen in receiving water and result in septic conditions
c. Floating materials such as grease or oil	Industrial wastes, especially from automobile industries	Interfere with treatment process and create toxic condition for ecosystem; may cause floating sludge/ scum problem
4. Biohazards		
Pathogenic bacteria	Domestic and hospital wastes	Transmit infectious diseases and may lead to epidemics

Different contaminants have different impacts on the environment, as presented in Table 4.4.

Because of urbanization and industrialization, a large quantity of water is required for industrial activities; only a small fraction of the water used is incorporated into their products and lost by evaporation, the rest finds its way into the water course as wastewater.

Industrial waste either joins streams or other natural water bodies directly, or is discharged into municipal sewers. Thus, such wastes affect in one way or another the normal aquatic life of a stream or the normal functioning of municipal wastewater treatment plants. Unlike domestic waste, industrial wastes are very difficult to generalize, as their characteristics not only vary with the type of industry but also from one plant to another plant producing the same type of end products. The examination of industrial wastewater is based on the type of industry and its product characteristics.

4.4.2. Quality Assessment

Table 4.5 represents the parameters essential for analysis to assess the efficiency of treatment processes. Also included are the parameters to be monitored in aeration basins to achieve maximum biological oxidation of organic wastes in an activated sludge treatment process.

TABLE 4.5
Parameters Relevant to the Treatment of Domestic Wastewater

Parameters	Raw wastewater	Secondary effluent	Tertiary effluent	Influent under aeration
Physical				
Color	X	X	X	
Odor	X	X	X	
Temperature	X		X	X
Turbidity	X	X	X	
Total suspended solids	X	X	X	X
Volatile suspended solids	X	X	X	X
Total dissolved solids	X	X	X	
Residue	X	X	X	
Sludge index				X
2. Chemical				
pH value	X	X	X	X
Alkalinity	X	X	X	

continued

TABLE 4.5 (continued)

Parameters	Raw wastewater	Secondary effluent	Tertiary effluent	Influent under aeration
Conductivity	X	X	X	
Dissolved oxygen	X			X
Organic carbon	X		X	X
Hardness	X	X	X	
3. Inorganics				
COD	X	X	X	
Ammonia	X	X	X	
Nitrate	X	X	X	
Nitrite	X	X	X	
Total nitrogen	X	X	X	
Chloride	X	X	X	
Residual chlorine			X	
Sulfate	X	X	X	
Sulfite	X	X	X	
Sulfide	X		X	
Fluoride	X	X	X	
Cyanide	X	X	X	
Phosphate	X	X	X	
4. Inorganic traces		All parameters should be analyzed where treated water is reused after secondary treatment only		
Arsenic	X		X	
Boron	X		X	
Cadmium	X		X	
Chromium	X		X	
Copper	X		X	
Mercury	X		X	
Nickel	X		X	
Silver	X		X	
Zinc	X		X	
Organics				
BOD	X	X	X	
Phenols	X	X	X	
Surfactants	X	X	X	
Grease/oil	X	X	X	

continued

TABLE 4.5 (continued)

Parameters	Raw wastewater	Secondary effluent	Tertiary effluent	Influent under aeration
6. Microbiology		All parameters should be analyzed where treated water is reused after secondary treatment only		
Total count	X		X	
Fecal indicators	X		X	
Fecal streptococci	X		X	
Salmonella	X		X	
Intestinal virus	X		X	
Multicellular organism	X		X	X
Protozoa				X

Table 4.6 describes the essential parameters to ascertain the suitability of industrial wastewater for discharge into the sea or sewerage networks.

TABLE 4.6
Essential Parameters to Ascertain the Suitability of Industrial Wastewater for Discharge into the Sea or Sewerage Networks

Parameters	Acceptability to the sea	Acceptability to sewerage networks
Physical		
Color	X	X
Odor	X	X
Temperature	X	X
Turbidity	X	X
Total suspended solids	X	X
Total dissolved solids	X	X
Residue	X	X
Chemical		
pH	X	X
Alkalinity	X	X
Conductivity	X	X
COD	X	X
Ammonia	X	X
Nitrate	X	X

TABLE 4.6 (continued)

Parameters	Acceptability to the sea	Acceptability to sewerage networks
Organic nitrogen	X	X
Total nitrogen	X	X
Sulfide	X	X
Cyanide	X	X
Hardness	X	X
Phosphate	X	X
Iron (Fe)	X	X
Copper (Cu)	X	X
Chloride		X
Chlorine demand		X
3. Inorganic traces		
Arsenic (As)	X	X
Boron (B)	X	X
Cadmium (Cd)	X	X
Chromium (Cr)	X	X
Mercury (Hg)	X	X
Nickel (Ni)	X	X
Silver (Ag)	X	X
Zinc (Zn)	X	X
Organics		
BOD	X	X
Phenols	X	X
Grease/oil	X	X
Detergents	X	X
Chlorinated hydrocarbon	X	X
Hydrocarbons	X	X
Microbiology		Examined in the wastes collected from hospitals, slaughterhouses, meat processing, tanneries, confectioneries, and other related industries.
Total coliforms		
Fecal coliform	X	
Pathogens	X	

Chapter 5

Water Quality Requirements and Standards

The utilization of water can be classified into five major categories:

1. Public supply for drinking and routine domestic activities
2. Recreational activities like swimming, fishing, skiing etc.
3. Preservation of aquatic life and wildlife
4. Agriculture
5. Industry

Water quality requirements differ according to the water's proposed use. Two main terms are associated with the assessment of the quality of water:

1. Water Quality Requirements

   This term represents a known or assumed necessity of water quality, which depends on the prior experience of the water consumer.
2. Water Quality standards

   These are the permissible levels of pollutants set up by a governmental or any international agency according to the proposed use of water. These limits are established to regulate the quality of receiving waters and to protect the public and the environment from the ill effects of harmful pollutants.

5.1 Surface Water

As explained in Section 4.1, surface water collected in ponds, lakes, rivers and the sea is a life support for an aquatic ecosystem, and hence must be be aesthetically pleasant and free from any sort of pollutant. Additionally, if it is needed as a source of drinking water, its quality should fall under drinking water standards and permissible limits because it will be directly associated with human health.

Normally, surface waters dissolve different salts and other soluble substances and even pick up soil bacteria and microbial life while flowing from higher regions to lower regions or to the collection points such as lakes, rivers etc. In addition to this, natural and synthetic fertilizers also enter surface waters along with agricultural wastes from irrigation sites. The situation worsens when domestic and industrial wastes are discharged to surface waters without any treatment. Hence, treatment of surface waters with conventional methods is essential to achieve adequate quality.

Water quality criteria, which apply to surface waters, are dependent on its intended use and perceived level of human contact. Generally, the following factors are taken into account if stream water is used for bathing, recreational activities and fishing:

- Water must be aesthetically pleasant.
- It must not contain any substance that is toxic on ingestion, or might produce irritation to the eyes or skin.
- It must be free from any sort of pathogenic or harmful microbes capable of causing infectious diseases and epidemics. Fecal coliform is considered to be the indicator organism to represent the extent of pollution of fecal origin in surface waters

The common factors to be examined to estimate the pollution level are temperature, color, biological oxygen demand (BOD), suspended solids, pH, ammonia, phosphorus, and heavy metals.

Oil and grease and indicator organisms are the common contaminants usually contributed to surface waters by discharges of treated or untreated effluents from municipal and industrial origin.

To protect human health, propagation of aquatic life and environments, international agencies have set up the stream water quality criteria, which consist of numerical limits of each pollutant. Some countries follow their own criteria set up by their individual governmental agencies.

The following tables describe the quality of stream water for bathing, swimming and maintenance of aquatic life in various regions such as the USA, Germany, Britain and the Middle East.

5.1.1 Stream Water Quality

The stream quality criteria set up by FRG (Federal Republic of Germany) and EPA (Environmental Protection Agency, USA) are presented in Table 5.1. Table 5.2 represents the water quality criteria for streams followed in Britain.

TABLE 5.1
Criteria for Stream Waters - German and EPA Guidelines

Parameters	Units	Minimum requirement (FRG)*	Purification required (FRG)*	EPA criteria**
Max. Temperature	°C			20
• cool summer water		25	25	
• warm summer water		28	28	
Oxygen	mg / l	4	4	5
pH	unit	6–9	6–9	6 . 5–8 . 5
Ammonium ($NH4^+$ - N)	mg / l	1	2	
BOD without nitrification inhibitions	mg / l	7	10	
COD	mg / l	20	30	
Phosphorus	mg / l	0.4	1	0.1
Iron	mg / l	2	3	—
Zinc	mg/l	1	1.5	—
Copper	mg/l	0.05	0.06	> 0.1
Chromium	mg/l	0.07	0.1	0.05
Nickel	mg/l	0.05	0.07	0.05

* Adapted from Rump, H.H. and Krist, H., *Laboratory Manual for the Examination of Water, Wastewater and Soil*, 2nd ed., VCH Verlagsgesellschaft mbH, with permission.

** Adapted from EPA, *Quality Criteria for Water*, U. S. Environmental Protection Agency, Office of Water Regulations and Standards, EPA 440 / 5 – 86 – 001, Washington D.C., 1976.

TABLE 5.2
Water Quality Criteria for British Streams

Quality class	Limits (observed 95% of the time)	Remarks	Use of stream water
1 A	Oxygen saturation > 80%; ammonium ($NH4^+$) ≤ 0.4 mg/L, BOD ≤ 3 mg/L; not toxic to fish	Average BOD ≤ 1.5 mg/L, no visible pollution	High quality water usable for drinking and other uses including contact recreation and fishing
1 B	Oxygen saturation > 60%; ammonium ($NH4^+$) ≤ 0.9 mg/L, BOD ≤ 5 mg/L; not toxic to fish	Average BOD ≤ 2 mg/L, $NH4^+$ ≤ 0.5 mg/L, no visible pollution	Water of lower quality than 1 A, but usable for the same purposes
2	Oxygen saturation > 40%; BOD ≤ 9 mg/L; not toxic to fish	Average BOD ≤ 5 mg/L, no visible pollution other than slight color because of humic compounds, little foam formation below dams	Usable for drinking after physical and chemical treatment, fishing possible for rough fish, moderate amenity value

continued

TABLE 5.2 (continued)

Quality class	Limits (observed 95% of the time)	Remarks	Use of stream water
3	Oxygen saturation > 20%; aerobic conditions, BOD ≤ 17 mg/L		Fish absent or only sporadically present; low-level industrial use; other uses only after treatment
4	Oxygen saturation worse than class 3, sometimes anaerobic		Waters that are grossly polluted and are likely to cause nuisance

Adapted from Natural Water Council, *Review of Discharge Consent Conditions*, London, 1978.

5.1.2 Quality of Bathing Water

The requirements for the quality of water suitable for recreational activities and bathing are given in Table 5.3. EPA and EEC (European Economic Community) set up these guidelines.

TABLE 5.3
Water Quality Requirements for Recreational Activities and Bathing

Parameters	Units	Normal value	Limits	EPA criteria*
Transparency	m	2	1	
pH	unit	-	6–9	6–9
Mineral oils	mg/l	- (< 0.3)	no visible film	free from oil/grease
Anionic surfactants	mg/l	- (< 0.3)	no foam formation	free
Phenol index	mg/l	0.005	0.005	-
Tar residue, suspended particles		none	-	-
Total coliform / 100 ml		500	1000	-
Fecal coliform / 100 ml		100	200	200
F. Streptococci /100 ml		100	-	-
Salmonella / 100 ml		-	0	0
Intestinal virus PFU / 10 ml		-	0	0

* Adapted from EPA, *Quality Criteria for Water*, U. S. Environmental Protection Agency, Office of Water Regulations and Standards, EPA 440 / 5 – 86 – 001, Washington DC, 1976.

5.1.3. Guidelines for Swimming Pool Water

The standards for water quality in swimming pools followed in USA and Middle Eastern countries region are mentioned in Table 5.4.

TABLE 5.4
Water Quality Standards for Swimming Pools Followed in USA and Middle East

Parameters	USA*	Middle East
Microbiology:		
a. Total count	When 10 ml portions of the samples are tested by MPN method — not more than 15% in any month should show agar plate count at 35°C of more than 200 colonies per ml	Membrane-filter method 10 to <100 bacterial count per 100 ml.
b. Total coliform	i) MPN — any five 10 ml portions not more than 2.2 MPN per 100 ml or ii) Membrane filter — 1 coliform count / 50 ml	nil
c. Fecal coliform	Free from all kinds of indicator and pathogenic bacteria	nil
d. Salmonella		nil
e. P. aeruginosa		nil
f. Vibrio cholerae		nil
g. Pathogenic parasites		nil
Free residual chlorine (ppm)	at least 0.4	0.5
Alkalinity (mg/L)	50	-
pH	7.0–8.2	6.5–8.5
Turbidity	nil	nil
Temperature (0°C)	not more than 30	-
Surfactants / Oil	free	free

* APHA, *Suggested Ordinances and Regulations Covering Public Swimming Pools*, American Public Health Association, 1984.

5.1.4. Water Quality Criteria for Aquatic Organisms

Water quality criteria for the protection of fish and other aquatic animals based on German and EPA guidelines are presented in Table 5.5.

TABLE 5.5
Stream Water Quality Requirements for Propagation of Aquatic Life

Parameters	Unit	Salmonid waters (FRG)[a]	Cyprinide waters (FRG)	Salmonid waters (EPA)[b]
Max. temperature	°C			based on geographical conditions, fish type and their resistance to temp. changes
• cool summer water		20	25	
• warm summer water		20	28	
Oxygen	mg/l	6	4	5 - fresh water, 6 - cold water, 4–5 - warm water

continued

TABLE 5.5 (continued)

Parameters	Unit	Salmonid waters (FRG)[a]	Cyprinide waters (FRG)	Salmonid waters (EPA)[b]
pH	unit		6.5–8.5	6.5–9.0 - fresh water 6.5–8.5 - marine water
Ammonium (NHr + -N)	mg/l	1	1	0.02
BOD without nitrification	mg/l	6	6	—
COD	mg/l	20	20	—
Iron	mg/l	2	2	<1
Zinc • with 4 mg/l Ca • with 20 mg/l Ca • with 40 mg/l Ca	mg/l	0.03 0.2 0.3	0.3 0.7 1	$<0.01 \times 96$ h-LC_{50}
Copper • with 4 mg/l Ca • with 20 mg/l Ca • with 40 mg/l Ca	mg/l	0.005 0.022 0.04	0.005 0.022 0.04	$<0.01 \times 96$ h - LC_{50}^{*}
Nitrite (NO2 - N)	mg/l	0.015	0.015	-
Residual chlorine	μg/l	—	—	2 - Salmonid & 10 - fresh or marine water organism
Phenol	μg/L			0.1
Grease / oil	mg/L			$<0.01 \times 96$ h - LC_{50}^{*}
Phosphorus	μg/L	—	—	0.1
Fecal coliforms / 100 ml				14 with not more than 10% of the samples exceeding 43/100 ml

[a] Adapted from Rump, H. H. and Krist, H. *Laboratory Manual for the Examination of Water, Wastewater and Soil*, 2nd ed., VCH Verlagsgesellschaft mbH with permission.

[b] Adapted from EPA, Quality Criteria for Water, U.S. Environmental Protection Agency, Office of Water Regulation and Standards, EPA 440/5–86–001, Washington D.C., 1976.

* $< 0.1 \times 96$h – LC50 – The concentration of a toxic compound that is lethal to 50% of the organism tested under the conditions in a 96h period of assay.

5.2 Ground Water

As explained in Section 4.2, the quality of ground water differs from one source to another. It is greatly affected by changes in the quality of waters entering the ground from different sources. Pollutants may enter ground water due to the entry of wastes rich in organic and inorganic pollutants in aquifers from municipal- and industrial-waste discharges. The penetration of organic pollutants is restricted because of their macromolecular structure, but inorganic pollutants more easily find their way to ground-water sources and cause contamination of ground waters. The main inorganic radicals are carbonates, sulfates and chlorides, which increase the hardness of water. On the other hand, ground waters absorb gases produced during bio-decomposition of organic matter present in waste discharges. The main gaseous constituent is CO_2, which lowers the pH. Others are H_2S and methane, the major cause of odor generation in water.

Thus, considerable care should be taken to protect ground-water storage from deterioration or contamination through the disposal of solid or liquid wastes. Since ground water may be utilized in the same way as surface water as a source of drinking water or for irrigation, surface water standards should be applied in the evaluation of its quality.

5.2.1 Quality of Ground Water

The quality of water obtained from the ground in the Middle East region is presented in Table 5.6.

TABLE 5.6
Quality of Ground Water in Middle East

Parameters	Units	Range
Temperature	°C	25–40
Conductitivy	µS/cm	5,000–7,000
pH	pH units	7.0–8.5
Alkalinity	mg/L	60–120
Hardness as $CaCO_3$	mg/L	1,500–2,000
Total dissolved solids, TDS	mg/L	3,500–5,000
Chloride	mg/L	1,000–1,500
Fluoride	mg/L	1.0–2.0
Ammonia	mg/L	0.01–1.0
Sulfate	mg/L	1,000–1,500
Silicates	mg/L	15–20
Calcium	mg/L	400–500
Magnesium	mg/L	100–200
Sodium	mg/L	500–650
Potassium	mg/L	15–30
Iron	mg/L	0.1–0.6
Manganese	mg/L	0.001–0.20
Aluminum	mg/L	0.02–0.06
Barium	mg/L	0.02–0.035
Strontium	mg/L	6.0–12.0

5.3 Potable Water

Potable or drinking water is considered the greatest beneficial use of water for mankind. It must be free of health hazards, i.e., pathogens, toxins, and carcinogens. Esthetic factors such as transparency, temperature, taste, odor, and chemical balance are also important in assessing the quality of potable water. Thus, standards are categorized as follows:

- Primary Standards: These standards are directly related to human health and describe limits of microbial contamination, inorganic and organic toxic chemicals, and turbidity. Prescribed limits must never be exceeded in water meant for drinking purposes.
- Secondary Standards: These represent the recommended contaminant levels for esthetic considerations such as inorganic chemicals, total dissolved solids, odor, corrosivity and color.

The standards for drinking water set up by the World Health Organization (WHO) and the EPA are included in this section. The potable water standards followed in the Middle East are also described.

5.3.1 WHO (World Health Organization) Standards

The quality of drinking water is regulated in most countries by recommendations or legal requirements. The recommendations of WHO, released in 1984, are of special importance for different countries. These recommendations are given in Table 5.7.

TABLE 5.7
WHO Recommendations for Drinking Water Quality

1 - Primary Standards - Microbiology Contaminants		
Bacteria	Limit (Count/100 ml)	Remarks
Treated water fed into mains		
1. *E. coli*	0	Turbidity < 1 NTU; pH -value on cholriantion < 8.0; after 30 min contact time free chlorine 0.2 - 0.5 mg/L
2. Coliforms	0	
Untreated water fed into mains		
1. *E. coli*	0	In 98% samples tested per year in larger supply systems
2. Coliforms	0	
3. Coliforms	3	Occasionally but not in sequential samples
Mains water		
1. *E. coli*	0	In 95% samples tested per year in larger supply systems
2. Coliforms	0	
3. Coliforms	3	Occasionally but not in sequential samples
Non-mains water supply		
1. *E. coli*	0	
2. Coliforms	10	should not appear in sequential samples

continued

TABLE 5.7 (continued)

2 - Primary Standards - Chemical Contaminants			
Inorganic constituents	Limit (mg/L)	Organic constituents	Limit (µg/L)
Turbidty (NTU)	5, (1 after disinfection)	Aldrin, dieldrin	0.03
		Benzene	10
Arsenic (As)	0.05	Benzo(a)pyrene	0.01
Cadmium (Cd)	0.005	Carbon tetrachloride	3
Chromium (Cr)	0.05	Chlordane	0.3
Cyanide (CN)	0.1	Chloroform	3
Fluoride (F)	1.5	DDT	1
Lead (Pb)	0.05	1,2 dichloroethene	10
Mercury (Hg)	0.001	1,1 dichloroethane	0.3
Nitrate (NO_3^-)	45	Trichloroethene	30
Selenium (Se)	0.01	Methaoxychlor	30
		Pentachlorophenol	10
		Tetra chloroethene	10
		2,4,6 - trichlorophenol	10
3 - Secondary Standards - Chemical Constituents			
Inorganic constituents	Limit (mg/L)	Inorganic constituents	Limit (µg/L)
Color (TCU)	15	Hydrogen sulfide	nil
Taste and odor	none	Sodium (Na)	200
pH	6.8–8.5	Iron (Fe)	0.3
Dissolved solids	1,000	Manganese (Mn)	0.1
Hardness as $CaCO_3$	500	Aluminum (Al)	0.2
Chloride	250	Copper (Cu)	1
Sulfate	400	Zinc (Zn)	5

Adapted from WHO, *Guideliens for Drinking Water Quality*, Vol. 1, Recommendations WHO, Geneva, 1984.

5.3.2 Potable Water Standards Established by EPA, USA

Drinking (potable) water must be tasteless and odorless, which means the dissolved solids must be in moderate quantities and the water free from suspended solids, turbidity, organic or toxic materials and pathogens. These EPA water regulations, established in 1975, are mentioned in Table 5.8.

5.3.3 Standards for Potable Water Followed in the Middle East

Standards for potable water followed in the Middle East region are presented in Table 5.9. Most of the permissible limits are fixed on the basis of WHO recommendations.

TABLE 5.8
Drinking Water Standards Set up by Environmental Protection Agency (EPA), USA

1 - Primary Standards - Microbiological Constitutents

Test method used	Monthly basis	Individual sample basis (less than 20 samples/month)	Individual sample basis (more than 20 samples/month)
Membrane filter technique	1/100 ml average density	Coliform count shall not exceed 4/100 ml in more than one sample	Coliform count shall not exceed 4/100 ml in more than 5% sample
MPN method			
1. Ten ml standard portions	Not more than 10% of the portions	Coliform shall not be present in more than 3 portions from one sample	Coliform shall not be present in more than 3 portions in 5% of the samples analyzed
2. The 100 ml standard portions	Not more than 60% of the portions	Coliform shall not be present in more than 5 portions from one sample	Coliform shall not be present in more than 5 portions in 20% of the samples analyzed

2 - Primary Standards - Turbidity Levels

Average	Maximum contaminate level (MCL) turbidity units (TU)
Monthly reading	1 TU or up to 5 TU (5 TU may be allowed provided it does not interfere with disinfection process, maintenance of chlorine residual or bacterialogical testing)
For two consecutive days' reading	5 TU

3 - Primary Standards - Inorganic Chenmicals

Contaminants	Level (mg/L)	Contaminants	Level (mg/L)
Arsenic (As)	0.05	Mercury (Hg)	0.002
Barium (Ba)	1.0	Nitrate (as N)	10.0
Cadmium (Cd)	0.010	Selenium (Se)	0.01
Chromium (Cr)	0.05	Silver (Ag)	0.05
Lead (Pb)	0.05		

4 - Primary Standards - Maximum Contaminant Level for Fluoride

Annual average of max. daily air temperature (°C)	Maximum contaminant level (MCL) (mg/L)
12.0 and below	2.4
12.1–14.6	2.2
14.7–17.6	2.0
17.7–21.4	1.8
21.5–26.2	1.6
26.3–32.5	1.4

TABLE 5.8 (continued)

5 - Primary Standards - Organic Chemicals	
Contaminants	Maximum contaminant level (MCL), mg/L
1. Chlorinated hydrocarbon	
Endrin	0.0002
Lindane	0.004
Methoxychor	0.1
Toxaphene	0.005
2. Chlorophenoxys	
2,4 - D (2,4 - dichloro phenoxy acetic acid)	0.1
2,4,5 - TP Silvex (2,4,5 - trichloro phenoxy propionic acid)	0.01
3. Trihalomethanes	0.10

6 - Secondary Standards - Limits for Esthetics			
Contaminants	Maximum contaminant level (MCL)	Contaminants	Maximum contaminant level (MCL)
Color	15 CU (color units)	Sulfate	250 mg/L
Odor	< 3 TON	Chloride	250 mg/L
Corrosivity	Noncorrosive	Copper	1..0 mg/L
Foaming agent	0.5 mg/L	Manganese	0.05 mg/L
pH	6.5–8.5	Iron	0.3 mg/L
Total Dissolved Solids	500 mg/L	Zinc	5.0 mg/L
Hydrogen sulfide	0.05 mg/l		

Adapted from EPA, *National Interim Primary Drinking Water Regulations*, U.S. Environmental Protection Agency, Office of Water Supply, EPA 570/9-76-003, 1976.

5.4 Wastewater

Wastewater collected from municipalities and communities must ultimately be returned to receiving waters or to the land after treatment. The treatment of wastewater is based on its composition, i.e., the actual amounts of physical, chemical and biological pollutants present in wastewater. Depending on the level of these pollutants, wastewater is classified as strong, medium or weak, as shown in Table 5.10.

5.4.1 Domestic Wastewater

The wastewater produced after domestic activities is treated by unit processes consisting of primary, secondary, and tertiary treatment processes. In primary treatment, physical operations such as screening and sedimentation are used to remove the

TABLE 5.9
Drinking Water Quality Standards in Middle East

1 - Chemical Constituents		
Parameters	**Unit**	**Requirement**
pH	pH unit	6.5–8.5
Turbidity	NTU	1
Conductivity	µS/cm	450–700
Total dissolved solids (TDS)	mg/L	250–450
Residual chlorine	mg/L	0.5–1.5
Dissolved sulfide	mg/L	nil
Total hardness	mg/L	70–150
Calcium hardness	mg/L	50–100
Alkalinity	mg/L	15–30
Chloride	mg/L	100
Ammonia	mg/L	nil
Iron	mg/L	0.1–1.0
Phenols	mg/L	nil
Chlorinated hydrocarbons	mg/L	As recommended by WHO
Trace elements	mg/L	Follow the same limits set up by WHO
2 - Microbiological Constituents		
Parameters	**Unit**	**Requirement**
Total count	count/100 ml	10 to < 100 (average value)
Total coliforms	count/100 ml	free from this bacteria
Pathogens • *E. coli* and *fecal streptococci* • *P. aeruginosa* • Parasites • Salmonella • Vibrio cholerae	count/100 ml	Free from all mentioned pathogenic species

floating and settleable solids found in wastewater. In secondary treatment, biological and chemical techniques are used to remove most of the organic matter. In tertiary treatment, additional combinations of unit operations and processes are used to remove other constituents, such as nitrogen and phosphorus, that are not reduced significantly by secondary treatment. Finally, the treated water is disinfected with any disinfectant such as chlorine to free it from any sort of pathogenic microbes. The normal concentration range of parameters maintained at various stages of treatment in the Middle East region is presented in Table 5.11.

5.4.2 Disposal of Wastewater into the Sea

After treatment, wastewater is either reused or disposed of in the environment, where it reenters the hydrological cycle. The fundamental consideration in

TABLE 5.10
General Composition of Raw Domestic Wastewater

Contaminants	Units	Weak	Medium	Strong
Total solids (TS)	mg/L	350	720	1200
Total dissolved solids (TDS)	mg/L	250	500	850
Fixed	mg/L	145	300	525
Volatile	mg/L	105	200	325
Total suspended solids (TSS)	mg/L	100	220	350
Fixed	mg/L	20	55	75
Volatile	mg/L	80	165	275
Settable solids	ml/L	5	10	20
Biological oxygen demand (BOD)	mg/L	110	220	400
Total organic carbon (TOC)	mg/L	80	160	290
Chemical oxygen demand (COD)	mg/L	250	500	1000
Nitrogen (as N)	mg/L	20	40	85
Free ammonia	mg/L	12	25	50
Organic nitrogen	mg/L	8	15	35
Nitrates	mg/L	0	0	0
Nitrites	mg/L	0	0	0
Phosphates (Total as P)	mg/L	4	8	15
Organic	mg/L	1	3	5
Inorganic	mg/L	3	5	10
Chlorides*	mg/L	30	50	100
Sulfates*	mg/L	20	30	50
Alkalinity (as $CaCO_3$)	mg/L	50	100	200
Grease / oil	mg/L	50	100	150
Total Coliforms	no./100 ml	10^6–10^7	10^7–10^8	10^7–10^9
Volatile organic compound (VOC)	µg/L	<100	100-400	>400

Adapted from Metcalf & Eddy, *Wastewater Engineering*, 3rd ed., McGraw-Hill, Inc., 1991. With permission.
* Values should be increased by amount present in domestic water supply.

discharging wastewater to the sea is its impact on the aquatic ecosystem. The emphasis should be given to the evaluation of parameters in discharges related to the dissolved oxygen level and toxicity such as suspended solids, pH, nutrients, pathogenic bacteria, and toxic chemicals including volatile organics, trace metals, pesticides, and polychlorinated hydrocarbons before discharging treated wastewater into the sea.

TABLE 5.11
The General Quality of Raw Domestic Wastewater, Secondary and Tertiary Effluents in Middle East

Parameter	Raw Wastewater	Secondary Effluent	Tertiary Effluent
Temperature (°C)	30–40	—	—
pH	6.0–8.0	6.0–8.0	6.0–8.0
Total solids	1,500–2,000	300–500	100–200
Total suspended solids (TSS)	200–300	20–30	10
Volatile suspended solids (VSS)	150–250	-	-
Chemical oxygen demand (COD)	600–900	-	-
Biological oxygen demand (BOD)	250–350	20–30	10
Sulfate (SO_4)	200–400	-	-
Sufilde (S^{2-})	5.0–20	-	0
Grease / oil	30–50	0	0
Residual chlorine	-	-	0.5–1.0
Ammonia (as N)	30–50	-	10–15

All the parameters except pH and temperatures are measured in mg/L.

Dissolved oxygen (DO) is an essential factor for the propagation of aquatic animals because detrimental effects can occur when DO levels drop below 4 to 5 mg/L depending on the aquatic species. Suspended solids affect water column turbidity and ultimately settle to the bottom, leading to possible benthic (organisms that may crawl, forming a burrow, or remaining attached to a substrate) enrichment, toxicity, and sediment oxygen demand. Nutrients can lead to eutrophication and DO depletion. The acidity of discharges measured by pH disturbs the chemical and ecological balance of receiving waters.

Toxic chemicals, including a range of organic and heavy metals present in discharges, affect aquatic life and human health. Consumption of such polluted waters or fish or other seafood propagating in these waters may cause diseases. Coliform bacteria are used as an indicator of pathogenic bacteria of fecal origin and as such provide a process to measure the safety of the water used for recreational activities and other human consumption purposes. Hence a criterion has been set up for the quality of treated wastewater discharges into the sea. Table 5.12 represents the quality of wastewater acceptable for discharge into the sea.

5.4.3 Reuse of Treated Effluent

Use of treated wastewater effluent as a reliable alternate source of water is receiving worldwide attention. Treated water provides both moisture and nutrients for the growth of plants, so can be successfully utilized for irrigation and landscaping activities.

TABLE 5.12
Quality of Wastewater Acceptable for Discharge into the Sea

Parameters	Unit	Limits
pH	pH unit	6.5–8.5
Total suspended solids (TSS)	mg/L	25
Total dissolved solids (TDS)	mg/L	1,000
Ammonia	mg/L	10
Dissolved sulfide	mg/L	5.0
Chlorides	mg/L	1,500
Free chlorine	mg/L	0.5
Phosphate	mg/L	5.0
Phenol	mg/L	0.05
Hydrocarbon	mg/L	25
Biological oxygen demand (BOD)	mg/L	10
Trace elements	mg/L	
Copper		1.5
Chromium		0.5
Mercury		0.01
Lead		0.5
Cadmium		0.05
Oil / Grease	mg/L	1.0

Countries that have hot, dry, desert-type climates usually practice the reuse of treated effluent for irrigation. The extremes of temperature and low humidity result in high rates of evaporation / transpiration in these countries, which means the water lost through evaporation from soil, surface water bodies, and transpiration from plants is very high.

Water required for irrigation can vary greatly in quality depending on the type and quantity of dissolved salts in the soil profile. This affects the characteristics of soil and crop yield.

A number of different guidelines for the quality of treated water utilized for irrigation have been set up. The general characteristics of treated effluent used for irrigation in the Middle East and permissible limits of each parameter set up by the Food and Agriculture Organization (FAO), USA are presented in Table 5.13.

5.4.4 Industrial Wastewater

Wastewater produced after industrial activities often contains many objectionable constituents, depending on the type of industry and its production unit. Hence, its quality must be monitored before it can be disposed of to natural waters or to municipal wastewater treatment plants.

TABLE 5.13
Characteristics of Treated Effluent Used for Irrigation in Middle East, and Permissible Limits Set Up by FAO

Parameters	Unit	Permissible limits set up by FAO*	Limits followed in Middle East
pH	pH unit	6.5–8.4	7.0–7.5
Conductivity	mho/cm	0.7–3.0	1.7–2.5
Specific ion toxicity			
Sodium (Na)	mg/L	-	175–420
Calcium (Ca)	mg/L	-	70–95
Magnesium (Mg)	mg/L	-	20–35
Sodium absorption ratio (SAR)		3.0–9.0	4.0–9.0
Chloride	mg/L	140–350	300–400
Boron	mg/L	0.7–3.0	0.2–1.0
Total nitrogen	mg/L	5.0–30.0	18–25
Hardness	mg/L	90–500	250–400
Residual chlorine	mg/L	1.0–5.0	0.5–1.0
Trace elements	mg/L		
Aluminum (Al)		5.0	2.0
Arsenic (As)		0.10	0.05
Cadmium (Cd)		0.01	0.005
Chromium (Cr)		0.1	0.05
Copper (Cu)		0.2	0.1
Iron (Fe)		5.0	0.5
Lead (Pb)		0.50	0.02
Manganese (Mn)		0.2	0.05
Mercury (Hg)		-	0.02
Nickel (Ni)		0.2	0.03
Zinc (Zn)		2.0	0.5

* Adapted from Metcalf & Eddy, *Wastewater Engineering*, 3rd Ed., McGraw-Hill, Inc., 1991. With permission.

The characteristics of industrial effluent acceptable for discharge to the municipal sewers and to the sea are mentioned in Tables 5.14 and 5.15 respectively.

TABLE 5.14
Quality of Industrial Effluent Acceptable for Discharge into the Municipal Sewers

Parameter	Units	Limits
Temperature	°C	Maximum 45
Color	TCU	Wastewater containing dyes should be discharged after decolorization
Ordor and taste	none	Should not cause nuisance
Toxicity		Should not affect the biological life in activated sludge process at wastewater treatement plant
Total suspended solids (TSS)	mg/L	300
pH	Units	6.0–9.0
Active chlorine	mg/L	0.5–3.0
Bromine	mg/L	1.0–3.0
Fluorides	mg/L	5.0
Ammonia nitrogen	mg/L	75
Nitrates	mg/L	As low as possible
Nitrites	mg/L	5.0
Total Kjeldahl nitrogen	mg/L	100
Cyanides	mg/L	0.1
Phosphorus (total)	mg/L	Should be kept as low as possible
Sulfate	mg/L	300
Sulfide	mg/L	0.5
Sulfite	mg/L	5.0
Chemical oxygen demand (COD)	mg/L	700–1,000
Biological oxygen demand (BOD)	mg/L	500
Chlorinated hydrocarbons	mg/L	0.1
Phenols	mg/L	5.0
Trace elements		
Aluminum (Al)	mg/L	20
Arsenic (As), lead (Pb), & boron (B)	mg/L	0.1 (each)
Cadmium (Cd)	mg/L	0.1
Chromium (Cr) & copper (Cu)	mg/L	1.0 (each)
Cobalt (Co)	mg/L	0.5
Nickel (Ni) & Zinc (Zn)	mg/L	2.0 (each)
Mercury (Hg) & silver (Ag)	mg/L	0.001 (each)

TABLE 5.15
Quality of Industrial Effluent Acceptable for Discharge into the Sea

Parameters	Unit	Limits
pH	pH unit	5.5–9.0
Temperature	°C	25–45
Total suspended solids (TSS)	mg/L	30
Total Kjeldahl nitrogen	mg/L	20
Ammonia - N	mg/L	10
Total nitrogen	mg/L	100
Hydrogen sulfide	mg/L	0.5
Cyanide	mg/L	0.5
Chemical oxygen demand (COD)	mg/L	200
Phenol	mg/L	1.0
Detergents	mg/L	1.0
Inorganic phosphorus	mg/L	0.5
Arsenic (As)	mg/L	0.1
Cadmium (Cd)	mg/L	0.005
Copper (Cu)	mg/L	0.1
Chromium (Cr) total	mg/L	0.3
Iron (Fe)	mg/L	2.0
Mercury (Hg)	mg/L	0.001
Manganese (Mn)	mg/L	0.2
Nickel (Ni)	mg/L	0.02
Lead (Pb)	mg/L	0.1
Selenium (Se)	mg/L	0.02
Vanadium (V)	mg/L	0.1
Zinc (Zn)	mg/L	2.0
Total coliforms	count/100 ml	1,000

Chapter 6

Determination of Physical Characteristics

6.1 Color — True and Apparent

Pure water is colorless, but foreign substances both in suspended and dissolved states impart color to water. Color imparted temporarily by suspended matter in water is called apparent color because it disappears on removal of suspended matter by filtration. Color imparted permanently to water, which remains even after removal of suspended matter, is known as true color. Hence, true color is due to the presence of colored dissolved solids in water.

The estimation of color is not included in routine domestic wastewater analysis. It should be done if domestic wastewater is receiving highly colored industrial wastes. In potable water, secondary- and tertiary-treated effluent and industrial water analysis measurement of true color is essential.

Sources

1. Organic and inorganic materials

a. Surface waters, mainly from swampy land, come into contact with wood and peat materials such as leaves, weeds, conifer needles, etc. These materials impart coloration to surface waters.
b. The decomposition of woody substances such as lignin produces tannins, humic acid, and humates, the colored byproducts. These components impart yellow coloration to water.
c. Surface waters attain red and brown coloration due to dissolution of iron and manganese oxides.

2. Industrial wastes

Significant coloration may be added to water streams by the discharge of untreated industrial wastes, e.g., from textile production, dyeing, pulp and paper, food processing, chemical, mining, refineries, slaughterhouses, tanneries, etc.

Significance

1. Palatability

Colored water is not palatable, that is, not aesthetically acceptable for public consumption. Its use without removal of coloration in various industries such as laundries, dyeing, paper, beverage, textile, food processing and plastic is not permitted. Thus, colored water affects the value of water for domestic and industrial uses.

2. Chlorine demand

Colored water may exert a chlorine demand and reduce the disinfecting power of chlorine when used in the treatment of water. Sometimes colored water produces objectionable odorous and colored organic compounds after reaction with chlorine. These may be toxic or carcinogenic in nature.

For these reasons, detection and removal of color from water prior to its use is essential in water management.

Measurement — Pt-Co Method

1. Principle

The true color of water and wastewater samples can be measured by the Pt-Co method, which uses Potassium Chloroplatinate (K_2PtCl_6) as the standard. One unit of true color of a centrifuged sample is equivalent to the color produced by 1 mg/L of Pt in the form of Chloroplatinate ion. The results are expressed as Pt-Co units of true color.

Note: *If the measurement of apparent color is required, do not centrifuge the sample.*

2. Interference

Any turbidity due to suspended matter may interfere with the result by producing false color reading. So the sample must be centrifuged before examination of true color.

3. Apparatus

a. Spectrophotometer — Hach DR/4000
b. Sample cells — two thoroughly washed
c. Volumetric flasks — with varying capacity

d. Pipettes
e. Centrifuge — any commercially available model with head capacity to hold four 50-ml centrifuge tubes.

4. Reagents

a. DDW: Color free
b. Standards: Hach Standard solution containing 500 Pt-Co units

5. Standardization

To perform color calibration, prepare calibration standard solutions containing 50, 100, 200, 300, and 400 Pt-Co units as follows:

a. Into five different 100-ml volumetric flasks, measure 10, 20, 40, 60, and 80 ml of standard solution containing 500 Pt-Co units.
b. Dilute each standard solution to the mark with DDW.
c. Pipette 10-ml DDW in a sample cell. This will represent the blank. Wipe the cell and keep in the cell holder. Set zero in the spectrophotometer.
d. Fill another sample cell with 10-ml standard solution and place in the cell holder after wiping with a soft clean cloth.
e. Set the wavelength at 465 nm.
f. Generate a standard curve following the instructions of the manufacturer.

6. Procedure

a. Centrifuge the sample to remove any kind of suspended matter causing turbidity.
b. Fill a sample cell with 10 ml of DDW to represent the blank. Wipe the cell, keep in the cell holder, and adjust zero.
c. Fill another sample cell with centrifuged sample up to 10-ml mark. Wipe the cells and keep in the cell holder.
d. Set the wavelength at 465 nm.
e. Record the reading directly and note it as Pt-Co units.

7. Calculation

Read directly from the readings of spectrophotometer or calculate the color units from the calibration curve.

8. Important Instructions

a. Clean the spectrophotometric cells as directed in Section 3.4.3.
b. Always wipe the cell with a soft clean cloth to prevent scratching the walls of the cell.
c. Always hold the cell by grasping its top. Stains on the cell wall obstruct the light path and interfere with the readings.
d. Always centrifuge the sample before determination of true color.

6.2 Temperature

This is one of the most essential parameters to determine in water and wastewater systems. It has significant impact on the growth and activity of ecological life. Also, it greatly affects the solubility of essential gases such as oxygen in water. Thus it is an important parameter to measure during the aeration of wastewater in the activated sludge treatment process.

Source

The temperature of natural water systems depends on many factors.

1. Weather conditions — i.e., atmospheric temperature

2. The subsequent discharge of cooling water in streams used for dissipation of waste heat in industries

3. Removal of forest canopies

4. Waste flows from irrigation sites.

Significance

1. Ecological Life

Functioning and propagation: Elevated temperature aggravates the metabolic activity of organisms present in water streams. Thus, the species that can utilize food efficiently and reproduce at high temperature may survive while the growth of others declines. At high temperatures some species may be eliminated completely. Accelerated growth of algae results at high temperature. Optimum temperature for normal activity and growth of common microbes is in the range of 25 to 35°C.

Aerobic digestion and nitrification cease when the temperature rises to 50°C. Methane-producing bacteria become inactive when temperature drops to 15°C, while autotrophic-nitrifying bacteria cease functioning at lower temperature (about 5°C).[7] Thus, variation in temperature of water and wastewater considerably affects the functioning and propagation of biological life.

2. Solubility of Gases

High temperature affects the solubility of gaseous components, mainly oxygen, which decreases at elevated temperatures. This, in turn, affects the growth and activity of aquatic organisms and ultimately influences the oxidation of organic matter present in streams, impoundment, wastewater-treatment plants, etc. In activated sludge treatment processes, the activity of aerobic bacteria, which are responsible for the oxidation of organic matter, is considerably reduced with an increase in temperature. Hence, this is an important parameter to measure in the influent of wastewater-treatment plants.

Measurement

1. Principle

The temperature of a water sample can be measured with the help of an ordinary mercury or digital thermometer.

2. Apparatus

a. Thermometer to measure air temperature — calibrated mercury thermometer with 0.5°C gradation, measurement range -20° to +60°C.
b. Thermometer to measure water and wastewater temperature — calibrated mercury thermometer with 0.1°C gradation, measurement range 0 to 100°C.
c. Sampling bottle — polyethylene, 5-L capacity

3. Procedure

a. Measure the air temperature with a dry thermometer approximately above the sampling point. Thermometer must be shaded from the sun.
b. Measure the water temperature by dipping the thermometer to the required depth in water or wastewater stream or aeration basins and wait until a constant reading is obtained.
c. Where direct measurement is not possible, collect 5 L water in sampling bottle. Immediately measure the temperature with thermometer.

6.3 Odor — Threshold Odor Number (TON) Method

Odor and taste are two closely associated terms. Both depend on the stimulation of a human receptor cell by the presence of some chemicals in water. Hence, these terms are oftenly included in "chemical senses." The substances producing odor in water mostly impart an objectionable taste as well. The reverse is not always true. There are many minerals known that impart bitter or salty taste to water without any contribution to odor production.

Sources

Natural waters come into contact with many chemical and biological substances available in the environment. Many such materials that impart odor and taste to water are added during human consumption of water for various activities. These substances are of three kinds.

1. Inorganic compounds

Alkaline salts of metals and various minerals mostly impart bitter or salty taste to water without any contribution to odor production. Hydrogen sulfide is one end product of the biological assimilation of organic matter. This imparts a "rotten egg" odor and unpleasant taste to water.

2. Organic compounds

Organic materials cause both odor and taste problems in water because of anaerobic biological decomposition of organic matter. The major components are hydrocarbons or petroleum-based products such as phenols, a variety of aromatic components, and also sulfur-based organics such as mercaptans.

3. Biological sources

Certain species of algae produce a malodorous oily substance that may result in both significant odor and unpleasant taste in water.

4. Wastewater discharges

Discharge of wastewater to natural waters is another source of odor in water. These are mainly due to the generation of malodorous volatile compounds such as hydrogen sulfide and ammonia. These are the end products of the anaerobic decomposition of sulfur- and nitrogen-containing organic compounds present in wastewater.

Significance

Odor and taste are recognized as essential parameters influencing the palatability of drinking water, and the tainting of fish and other aquatic organisms and aesthetics of recreational water. Consumers find odor and taste aesthetically unpleasant for obvious reasons. Such waters also pose a health threat because of the presence of decomposed organic constituents. These may be toxic or carcinogenic, for example, chlorophenol. This has been found to produce an unpleasant taste in fish at a concentration of only 0.0001 mg/L. Similarly, copper at concentration of 0.019 mg/L imparts a green color and characteristic unpleasant taste to oysters.[8] Exposure to the offensive odors of sulfur-containing compounds present in wastewater can cause poor appetite, lower water consumption, impaired respiration, nausea and vomiting, and sometimes mental perturbation. Long exposure to 300-ppm hydrogen sulfide in air may cause death.[9]

Characterization of Odor

The odor present in water and wastewater may be health hazardous or toxic or may cause psychological stress. Hence, the complete characterization of odors is an important aspect of water and wastewater analysis. This depends on four independent factors: noticeable features, threshold level, palatability, and strength as described in Table 6.1.

TABLE 6.1
Factors Essential for Complete Characterization of Odor

Factors	Description
Noticeable features	Described by the subject on exposure to the odors depending on their olfactory capability
Palatability	Pleasant or unpleasant as sensed by the subject
Threshold level	The minimal detectable odor concentration achieved after dilutions
Strength	The intensity of the odor. Usually measured by dilution method (TON) or butanol olfactometer.

Measurement

1. Principle

The malodorous compounds responsible for producing objectionable odors in water can be detected by diluting a sample with odor-free water until the least detectable odor level is achieved. This is recorded as TON (Threshold Odor Number). The concentration of malodorous gases such as hydrogen sulfide, ammonia, mercaptans etc. emitted into the air from wastewater can be measured by any commercially available gas monitor.

2. Apparatus

a. Sampling bottles — Pyrex glass, narrow-mouth bottles with glass stoppers
b. Dilution flasks — Erlenmeyer flasks of 250-ml capacity with glass stoppers
c. Measuring cylinders of different capacities
d. Pipettes — 10-ml capacity

3. Reagents

Dilution water must be free from any sort of odor. DDW is used for this purpose.

4. Procedure

a. Varying quantities of odorous samples are measured into separate dilution flasks as mentioned in Table 6.2.
b. The sample in each flask is diluted to 200-ml with DDW.
c. A blank is prepared by measuring 200-ml DDW into a dilution flask.
d. An assembled panel of five to ten "noses" is used to determine the mixture in which the odor is just barely detectable to the smell and to record the results.

Note: "Noses" means testers, the persons selected for odor measurement.

TABLE 6.2
Threshold Odor Number (TON) corresponding to different dilutions

Sample Volume (ml)	TON	Sample Volume (ml)	TON
200	1.0	40	5.0
175	1.1	35	6.0
150	1.3	25	8.0
140	1.4	10	20
125	1.6	8.3	24
100	2.0	5.7	35
75	2.7	4.0	50
70	3.0	2.0	100
50	4.0	1.0	200

Adapted from Peavy, H. S., Rowe, D. R., and Tchobanoglous, G., *Environmental Engineering*, McGraw-Hill, Inc., 1985, with permission.

5. Calculation

The extent of odor present in sample can be calculated by the following relation:

$$TON = A + B / A$$

where A = Volume of odorous water sample, ml
B = Volume of odor-free water to make up final volume 200 ml, ml

Threshold odor numbers corresponding to various sample volumes are presented in Table 6.2.

Note: *Concentration of odors in water and wastewater can also be measured by sensitive olfactometers available in the market. Odors imparted by specific organic compounds can be successfully measured by gas chromatography.*

6. Important Instructions

a. The room where the odor test is performed should be free from any kind of odor such as smoke, perfume, shaving lotion etc.
b. The room should be properly ventilated.
c. The testers should not be suffering from colds or any sort of allergies.
d. The testers should not be allowed to prepare the samples.

6.4 Taste — Taste Threshold Method

For sources and significance, see Section 6.3.

Measurement

1. Principle

The quantitative measurement of taste is performed as explained for odor in Section 6.3. The results are presented as taste threshold.

2. Apparatus

The following additional apparatus are required along with those mentioned in Section 6.3:

a. Beakers — 50-ml capacity
b. Water bath — set at 40 ± 1°C.

3. Reagents

Dilution water: It must be tasteless. DDW is utilized for this purpose.

4. Procedure

a. Follow the dilution method as described in Section 6.3.
b. Transfer 30 ml of each diluted sample and blank (DDW) in separate clean 50-ml beakers.
c. Maintain the temperature of samples and blanks at 40 ± 1°C by keeping them in a water bath.
d. Tester should take a suitable volume of sample into mouth and hold for several seconds, then discharge without swallowing.
e. Record the comparison of taste of sample with blank.

5. Calculation

Calculate threshold values as described in Section 6.3.

6.5. Turbidity — Formazin Attenuation Method — Detectable Range: 0 to 100 FAU/NTU

Turbidity indicates the light-transmitting capability of water and wastewater with respect to colloidal and suspended matter. It is a measure of the extent to which light is either absorbed or scattered by suspended matter in water, but it is not a direct quantitative measurement of suspended solids.

Sources

The presence of turbidity in water is due to a variety of reasons.

1. Erosion of soil

Colloidal matter such as clay, silt, rock fragments, sand particles, and metal oxides present in the soil find their way to surface waters through soil erosion. These materials remain suspended in water and cause turbidity.

2. Discharge of wastewater

Wastewater generated after domestic and industrial activities when discharged to water streams adds significant quantities of organic and inorganic materials. Nutrients rich in N and P contents aggravate the growth of algae.[4] This is also another source of turbidity in water.

Industrial wastes containing soaps, detergents, oil/grease, and other emulsifying agents produce stable colloids that result in turbidity in water.

3. Microbiological propagation

Organic matter reaching surface waters provides enough nourishment for the growth of various microorganisms. This may result in additional turbidity in water. Vegetable fibers also contribute to turbidity.

Turbidity measurements are not commonly run on raw wastewater but are measured in treated effluent.

Significance

Turbidity measurement is an important factor related to the quality of public water supply. It should be measured in treated wastewater effluent if it is reused.

1. Public acceptability

Turbid water gives an aesthetically unpleasant opaqueness to water. The colloidal particles may serve as suitable adsorption sites for different chemicals. These can impart undesirable taste and odors or may be toxic or hazardous to health.

2. Anaerobiosis

The discharge of turbid water to natural streams can cause accumulation of sediments that could develop an anaerobic environment. Anaerobiosis results in the production of malodorous components that may adversely affect the flora and fauna of the stream.

3. Filtration

Filtration of highly turbid waters is an expensive process. The colloidal particles choke the sand filters, which results in a short filter run and increases the cleaning costs. This, in turn, results in the production of inferior-quality water and also interferes with reuse of treated effluents.

4. Disinfection

The colloidal particles available in turbid water make the task of disinfection more difficult and expensive. Due to adsorption capability, the colloidal particles may behave as a protective shield for harmful microbes such as pathogens.

5. Light penetration

The colloidal particles in turbid waters may interrupt the process of photosynthesis in the stream because these particles restrict the penetration of light into water.

Measurement

1. Principle

Turbidity is an optical property that results from the scattering and absorption of light by colloidal or suspended particles present in the sample. The intensity of light scattered by the sample is compared with the intensity of light scattered by a standard solution. Formazine solutions are used as standards for calibration.

2. Interference

a. The sample should be free from debris and rapidly settling particles.
b. Air bubbles interfere at all levels. Samples with foaming quality should be degassed before estimation using a degassing kit (Hach Company).
c. Color due to dissolved substances can impact the reading.
d. Dirty glassware or scratches on the wall of spectrophotometric cell obstruct the path of the light ray.

3. Apparatus

a. Spectrophotometer — Hach DR / 4000
b. Sample cells – two thoroughly cleaned spectrophotometric cells
c. Volumetric flasks
d. Pipettes

4. Reagents

a. Turbidity-free water: Use DDW. If the turbidity of water is greater than 0.05 FAU, pass it through a 0.45 μm membrane filter to remove any sort of turbidity.
b. Turbidity Standards: Purchase Formazin stock solution with 100 NTU or FAU from the manufacturer.

5. Standardization

For turbidity calibration, prepare Turbidity calibration standards containing 1.0, 10, 20, 50 and 100 FAU as follows:

a. Into four different volumetric flasks measure 1.0, 10, 20 and 50 ml of well-mixed Formazin stock standard containing 100 FAU.

b. Dilute each standard solution with DDW to 100 ml. Stopper and invert each flask to mix gently.

Note: *Do not shake the flasks. Shaking causes excessive froth and interferes with the standardization procedure.*

c. For 100 FAU concentration take the standard solution directly without dilution.

d. Fill the sample cell with standard solution and place in the cell holder.

e. Set the wavelength at 860 nm.

f. Fill another sample cell with DDW and keep in the cell holder. This will serve as a blank.

g. Record the absorbance value of each standard solution against the blank.

h. Generate a standard curve of absorbance against turbidity units in FAU or NTU.

6. Procedure

a. Fill the sample cell to 10-ml mark with DDW. Wipe the sides of sample cell with a clean soft cloth. This represents the blank.

b. Hold the sample cell by grasping its top.

c. Place the blank into the cell holder and set zero.

d. Mix an unfiltered sample thoroughly to disperse solids. When air bubbles subside, pour the sample into the sample cell up to 10 ml mark

e. Wipe it and keep in cell holder.

f. Record the reading directly at 860 nm.

Note: *Dilute the highly turbid waters and record the dilution factor.*

7. Calculation

a. The results are reported in FAU or NTU.

b. One Formazine Attenuation Unit (FAU) is equivalent to one Nephlometric Turbidity Unit (NTU).

If the sample is diluted, then:

$$\text{FAU or NTU} = A\,(B + C)/C$$

where A = FAU or NTU reading
B = Volume of DDW, ml
C = Volume of sample, ml

6.6 Residue

The term "residue" refers to solid matter suspended or dissolved in water or wastewater. Suspended and dissolved solids correspond to non-filterable residue and filterable residue respectively.

Source

A variety of solids in suspended and dissolved states are present in water and wastewater. They are of the following types:

1. Inorganic Solids:

The particles of clay, silt, and other soil components resulting from soil erosion remain suspended in surface waters. Inorganic substances such as minerals, metals, and gases constitute the dissolved solid fraction of water. Water may come in contact with these materials in the atmosphere, on land surfaces and within the soil.

2. Organic solids:

Organic materials such as vegetation, plant fibers, leaves, human / animal wastes etc. remain suspended in surface waters due to soil erosion. Organic suspended materials include bacterial population also. Volatile suspended solids (VSS) represent this fraction of residue.

The end products of biological decay of organic matter remain dissolved in water. This is represented in dissolved solid content of water and wastewater.

3. Grease or oil:

Immiscible liquids such as oils and grease remain suspended in water and are included in suspended-solid measurement.

All the above-mentioned solids may result from the use of water in domestic and industrial activities.

Significance

Water having suspended material is not acceptable for the following reasons:

1. The biological degradation of suspended organic solids may result in toxic, health hazardous, and malodorous by-products.
2. Microbial suspended solids may include pathogenic microbes causing diseases or may include toxin-producing, filamentous organisms.
3. Dissolved matter may impart unpleasant odor or taste to water. Some chemicals may be toxic or carcinogenic in nature, especially those produced by the action of chlorine with organic constituents such as chlorophenols.

Measurement

Figure 6.1 represents the sequence of the measurement of solids in water and wastewater.

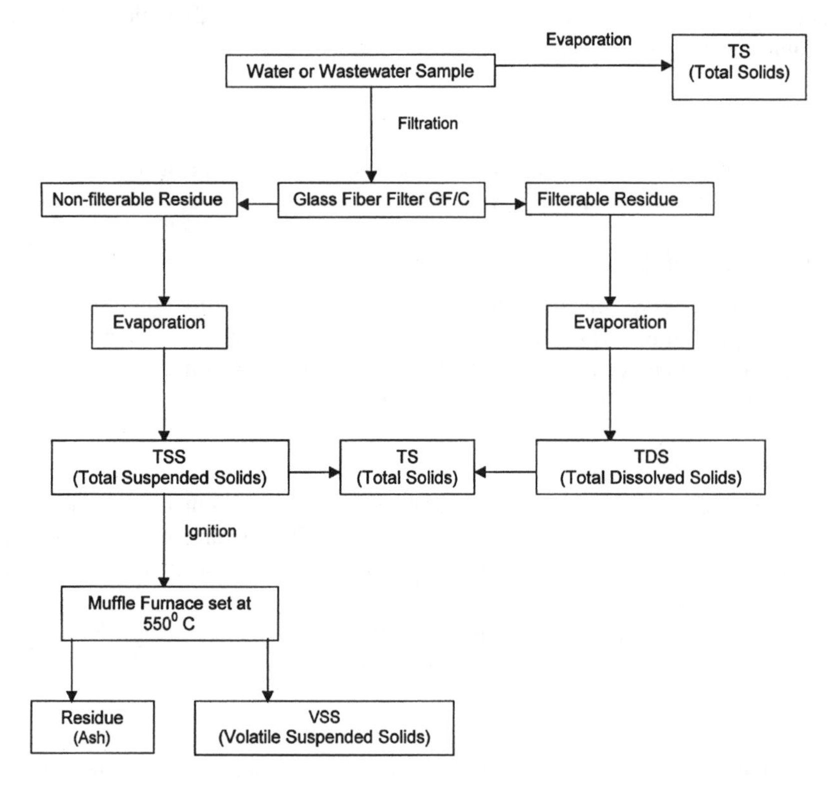

Figure 6.1 Schematic representation of interelationship of solids found in water and wastewater.

6.6.1 Nonfilterable Residue — Suspended and Volatile

TSS — Total Suspended Solids

1. Principle

The residues retained on a glass fiber filter after filtration are known as total suspended solids (TSS). The filter and solids are dried at 103 to 105°C to a constant weight.

2. Apparatus

a. Filtration unit — comprising a filtration flask of 1L capacity, a funnel, a filter holder, and a vacuum pump
b. Filter paper — Whatman GF/C or equivalent 4.25-cm disc size

c. Oven — set at 103°C
d. Desiccating cabinet
e. Forceps — flat-bladed without serrated tips
f. Analytical balance

3. Procedure

a. Keep the filter paper in an oven set at 103°C for 2 h. Cool to room temperature in a desiccating cabinet.
b. Weigh it to a constant weight with the help of an analytical balance.
c. Set up the filtration unit and place the preweighed filter paper on a filter holder.
d. Filter 100 ml of water or wastewater sample through the filter under vacuum. Rinse down the funnel with a small amount of DDW.
e. Place the filter paper at 103°C in an oven. Dry it to a constant weight. Usually leave the filter paper overnight in the oven.
f. Cool the filter paper in a desiccating cabinet and weigh to a constant weight. Record the weight.
g. Preserve the solids on filter paper for the measurement of volatile suspended solids.
h. Preserve the filtrate for the estimation of total dissolved solids.

4. Calculation

$$\text{mg TSS/L} = \frac{(A - B) \times 1000}{\text{volume of sample, ml}}$$

where A = weight of filter paper and solids, g
B = weight of filter paper, g

VSS — Volatile Suspended Solids

1. Principle

The volatile components of the nonfilterable residue are determined by igniting the filter and solids at 550°C in a muffle furnace after determining the suspended solids. The loss of weight after ignition represents volatile suspended solids (VSS).

Note: *The organic fraction of suspended solids will change to CO_2 at ignition temperature, leaving behind the inorganic fraction as "ash." At 550°C, the decomposition of inorganic fraction is least. Most of the inorganic salts are relatively stable, with the exception of $MgCO_3$. This decomposes at 350°C according to the following equation:*

$$MgCO_3 \rightarrow MgO + CO_2\uparrow \tag{6.1}$$

2. Apparatus

All apparatus required for TSS are required plus the following:

a. Muffle furnace — set at 550°C
b. Crucible

3. Procedure

a. Weigh a clean, dry crucible to a constant weight.
b. After measuring TSS, place filter paper along with solids in a preweighed crucible and ignite in a muffle furnace that has been preheated to 550°C. The process of ignition usually takes 20 to 30 min. to complete.
c. Allow the crucible to cool to room temperature by keeping it in a desiccator cabinet and weigh until a constant weight is achieved. Otherwise, repeat the process as mentioned above.

4. Calculation

$$\text{Mg VSS/L} = \frac{(A - B) \times 1000}{\text{Volume Sample, ml}}$$

where A = weight of filter paper and solids before ignition, g
B = weight of filter paper and solids after ignition, g

6.6.2 Filterable Residue — Dissolved and Total

TDS — Total Dissolved Solids

1. Principle

A filtered sample containing the dissolved solids is evaporated to dryness at 180°C. The residue is known as total dissolved solids (TDS).

2. Apparatus

Same as mentioned above, plus an evaporating dish.

3. Procedure

The estimation is carried out with the filtrate collected after the filtration of the sample containing suspended solids.

a. Take an evaporating dish cleaned with chromic acid and rinsed well with tap water, then with DDW. Dry it overnight in an oven set at 180°C.

b. Cool the dish in a desiccator cabinet to room temperature. Weigh until a constant weight is achieved.

c. In a preweighed evaporating dish, place 100-ml filtrate collected in the estimation of TSS.

d. Place it in the oven set at 180°C until all the filtrate evaporates, leaving behind the dissolved solids.

e. Cool in a desiccator cabinet and weigh the dish. Repeat the steps of cooling and weighing until a constant weight is achieved. Record the final constant weight.

4. Calculation

$$\text{mg TDS/L} = \frac{(A - B) \times 1000}{\text{Volume of Sample, ml}}$$

where A = weight of dish and filtrate, g
B = weight of dish, g

Total Solids — TS

1. Principle

An unfiltered sample containing both suspended and dissolved solids is evaporated to dryness at 103°C. The residue thus obtained is known as total solids (TS).

2. Apparatus

Same as described for TSS.

3. Procedure

a. Follow the procedure described above for TDS.

b. Take 50-ml sample directly without filtration.

c. Keep the drying temperature at 103°C.

4. Calculation

$$\text{mg TS/L} = \frac{(A - B) \times 1000}{\text{Volume of Sample, ml}}$$

where A = weight of dish and sample (without filtration), g
B = weight of dish, g

5. Important Instructions

a. The process of drying, cooling, desiccating, and weighing of filter paper or crucible should be repeated until a constant weight is achieved or weight loss is less than 4% of previous reading.

b. Always keep crucible in a muffle furnace that has been preheated to 550°C.

Chapter 7

Determination of Inorganic and Nonmetallic Constituents

7.1 pH — Electrometric Method

pH is a logarithmic scale generally used to express the acidic, alkaline, or neutral nature of a solution. In fact, it presents the hydrogen ion concentration, or, more precisely, the H^+ ion activity in a given solution.

Source

pH value is the best indication of the presence of acid or alkali in the water sample. Due to hydrolysis of dissolved salts, the pH value can decrease or increase beyond neutral value, i.e., 7.0, showing the presence of acid or alkali in the solution. The presence of salts of strong base and weak acid, e.g., Na_2CO_3 increases pH value; salts of weak base and strong acid, e.g., $CaCl_2$ decrease pH level. Thus, a fundamental relationship exists among pH, acidity, and alkalinity.

Significance

pH is an essential factor to be estimated in each and every phase of water and wastewater treatment.

1. Water

All the processes involved in the treatment of potable water, such as chemical coagulation, disinfection, softening, and corrosion control, are pH dependent.

2. *Wastewater*

a. The biological treatment of wastewater involves decomposition of organic matter available in wastewater by different species of aerobic bacteria. The growth and activity of these microbes depend on the pH level in wastewater.

b. Generation and emission of malodorous gases are also controlled by pH variations.

c. In chemical treatment of wastewater, the coagulation of wastewater, dewatering of sludge, and oxidation of certain substances such as cyanide are also pH-dependent processes.

Hence, accurate measurement and monitoring of this factor in optimum range is of great significance in water and wastewater management and treatment.

Measurement

1. *Principle*

On the basis of the Ionization Theory, acids are those substances that yield hydrogen (H^+) ions, and bases are those that yield hydroxyl (OH^-) ions on dissociation. According to this concept, strong acids and bases are highly ionized substances and weak acids and bases are poorly ionized substances in aqueous solutions.

To measure H^+ ion concentration, consider the dissociation of pure water at 25°C, which yields one OH^- ion for each H^+ ion.

$$H_2O \leftrightarrow H^+ + OH^- \tag{7.1}$$

According to the Law of Mass Action, the dissociation constant (k) at equilibrium will be

$$K = \frac{[H^+][OH^-]}{[H_2O]} \tag{7.2}$$

where the brackets represent the concentration of ionization products in moles/L. Since the concentration of water is extremely large and is slightly affected by the ionization process, it is considered as constant. Equation 7.2 can be rewritten as

$$k = [H^+][OH^-] \tag{7.3}$$

Now k is represented as K_w, which is known as ionic product or ionization constant for water at 25°C. Thus, Equation 7.3 is modified as

$$K_w = [H^+][OH^-] \tag{7.4}$$

It is seen from conductivity measurement that pure water is a very weak electrolyte, which on dissociation at 25°C furnishes only 10^{-7} mole H^+ ion per liter. Since water produces one OH^- ion for each H^+ ion on dissociation, the concentration of $[OH^-]$ ions is equal to $[H^+]$ ions concentration, i.e., 10^{-7} mole/L. Thus, for pure water at 25°C,

$$[H^+] = [OH^-] = 10^{-7}$$

therefore,

$$K_w = 10^{-7} \times 10^{-7} = 10^{-14} \quad (7.5)$$

If an acid is added to water it ionizes and releases more H^+ ions into the solution, the concentration of H^+ ions becomes more than 1×10^{-7} mole/L, and the solution is said to be acidic. Similarly, if a base is added to water it furnishes OH^- ions on ionization, which increases the concentration of OH^- ions to more than 1×10^{-7} mole/L. This solution is said to be basic or alkaline.

Thus, the concept of ionic product of water enables the solutions to be classified as acidic, neutral, or alkaline by considering the H^+ or OH^- ion concentration.

The concentration of H^+ or OH^- ions can never be reduced to zero, no matter how acidic or basic the solution may be.

In 1909, Sorensen introduced a logarithmic scale to determine the H^+ ion concentration that was named the pH Scale. According to this scale, the pH of a solution is defined as the negative logarithm of its H^+ ion concentration in moles/L.

Thus,

$$pH = -\log[H^+] \text{ or } pH = \log \frac{1}{[H^+]} \quad (7.6)$$

Equation 7.4 is modified as follows:

$$-\log K_w = -\log[H^+] - \log[OH^-]$$

or

$$p\,K_w = pH + pOH \quad (7.7)$$

where p stands for “power” or exponent.

Since, for pure water at 25°C,

$$K_w = 1 \times 10^{-14}, \text{ so } pK_w = 14$$

$$pH + pOH = 14 \tag{7.8}$$

On this basis, a pH scale has been developed ranging from 0 to 14. A solution having pH equal to 7 is neutral, the one having a pH value > 7 is acidic, and one having pH $< 7 \geq 14$ is basic in nature.

The determination of pH in water and wastewater samples is based on the determination of $[H^+]$ ion concentration by a potentiometer using a glass electrode and a reference electrode.

2. Apparatus

a. pH meter: Any pH meter available (manufactured by a reputed international manufacturing company such as ORION) consisting of reference and glass electrodes with automatic temperature control system.

b. Beakers

c. Stirrer: Magnetic stirrer with a magnetic, TFE-coated stirring rod.

3. Reagents: Buffer solutions

a. Acetate buffer, pH 4.62:

i. Acetic acid, CH_3COOH, 1M: Measure 16 ml, 94.5% acetic acid in a 250-ml volumetric flask. Make up the final volume to 100 ml with DDW.

ii. Sodium hydroxide, NaOH, 1M: Weigh 40 g NaOH and transfer to a 1L volumetric flask. Dissolve in DDW and bring the final volume to 1L with DDW. Cool the solution.

Measure 200-ml acetic acid in a 1L volumetric flask. Add 100 ml NaOH solution. Bring the final volume up to the mark with DDW. This is an acetate buffer of 4.62 pH.

b. Phosphate buffer, pH 7.0:

i. Potassium dihydrogen phosphate, KH_2PO_4: Weigh 9.078 g KH_2PO_4 and transfer to a 1L volumetric flask. Dissolve in DDW. Bring the final volume up to the mark with DDW.

ii. Disodium hydrogen phosphate, Na_2HPO_4: Weigh 11.88 g Na_2HPO_4 and transfer to a 1L volumetric flask. Dissolve in DDW. Bring the final volume up to the mark with DDW.

Measure 200 ml of KH_2PO_4 solution in a 500-ml volumetric flask. Add 300 ml of Na_2HPO_4 solution. The pH of this buffer will be 7.0.

c. Borate buffer, pH 9.0:

i. Boric acid, H_3BO_3: Weigh 12.4-g boric acid and transfer to a 1L volumetric flask. Dissolve in DDW. Add 100 ml of 1M NaOH solution (prepared for acetate buffer) into the flask. Bring the final volume to 1L with DDW.

ii. Hydrochloric acid, HCl, 0.1M: Measure 500 ml DDW in a 1L volumetric flask. Now add 8.3 ml concentrated HCl. Dilute up to the mark with DDW.

Measure 850-ml boric acid solution in a 1L volumetric flask. Add 150 ml, 0.1M HCl into it. The final volume should be 1L. This buffer has pH 9.0.

4. Calibration

a. Follow manufacturer's instructions for calibration or use of the instrument and for storage of the electrodes.

b. Before use, remove the electrode from storage solution and rinse with DDW. Dry electrode by gently blotting with a soft tissue.

c. Instrument should be set at room temperature, normally at 25°C.

d. Place electrode into the first buffer with pH 4.62.

e. Adjust the pH value in the meter.

f. Rinse electrode again with DDW and wipe gently.

g. Place into second buffer with pH 9.0. Wait for a stable display. The reading should be within ± 0.1 unit for the pH of second buffer.

Note: *If the difference is greater than ±0.1 unit from expected value, check the electrode or trouble shootings given in the instructions.*

h. Now the instrument is calibrated and ready for use.

5. Procedure

a. After calibration, rinse the electrode with DDW and wipe as mentioned above.

b. Take the sample in a beaker. Bring the temperature of the sample to room temperature.

c. Place the beaker on a magnetic stirrer. Bring the sample to homogeneity by stirring.

d. Place the electrode in the sample in such a way that the bar magnet will not touch the electrode.

e. Record the reading, which will give the pH value of the sample.

6. Calculation

The read-out of the pH meter will give direct pH value of the sample.

7.2 Alkalinity — Titrimetric Method

Alkalinity is defined as the capability of water and wastewater to neutralize H^+ ions. In other words, it is the ability of water and wastewater to accept protons or neutralize acids. It is measured by titration with acid.

Source

1. Atmosphere and soil

Alkalinity in water and wastewater is mainly due to the presence of hydroxide (OH^-), carbonate (CO_3^{2-}); bicarbonate (HCO_3^-) of metals like Mg, Ca, Na and K. Phosphates (HPO_4^{2-} or $H_2PO_4^-$), borates, silicates, HS^- (hydrogen sulfide) ions and ammonia would also contribute alkalinity. These compounds are present in water due to the dissolution of mineral salts available in the soil and atmosphere. Besides these, the major ionic components imparting alkalinity to water are HCO_3^-, CO_3^{2-}, and OH^-. The following general reactions produce these components:

$$CO_2 + H_2O \leftrightarrow H_2CO_3 \tag{7.9}$$

$$H_2CO_3 \leftrightarrow H^+ + HCO_3^- \tag{7.10}$$

$$HCO_3^- \leftrightarrow H^+ + CO_3^{2-} \tag{7.11}$$

$$CO_3^{2-} + H_2O \leftrightarrow HCO_3^- + OH^- \tag{7.12}$$

2. Industrial wastes

Phosphates are mainly added from effluent of industrial wastes rich in soaps and detergents. Fertilizers and pesticides present in agricultural wastes also contribute to phosphate alkalinity in water.

3. Biological sources

a. H_2S and NH_3 are the major products of microbial decomposition of organic material containing sulfur and nitrogen under anaerobic environment. They may impart alkalinity to water streams also.

b. Surface waters where algae are flourishing contain an appreciable amount of alkalinity because of excessive CO_2 dissolved in water. This reaction produces carbonic acid as shown in Equation 7.9.

4. Boiler waters

Boiler waters also contain high alkalinity due to the presence of CO_3^{2-} and OH^- ions.

5. Water softening

If lime or lime-ash is used as a softening agent during chemical treatment of water, it increases the level of alkalinity. Such samples, when discharged to water streams, impart alkalinity to water.

Significance

1. Palatability

In general, a high level of alkalinity imparts a bitter taste to water. Thus such water is not acceptable for public consumption, especially for drinking.

2. Water supply system

Alkaline water combines with certain cations present in water and precipitates them, which can affect pipes and other water supply equipment.

3. Treatment of wastewater

Alkalinity plays an important role in the treatment of wastewater, as it indicates the buffer capacity of water. This appreciably affects the growth and activity of microbes present in activated sludge, which are responsible for the treatment of wastewater.

It is also an essential parameter to be estimated to design and implement the corrosion- and odor-control processes.

4. Purification of water

In order to design and operate a chemical treatment process for the purification of water, alkalinity measurement is important to select the appropriate chemical for coagulation and softening of water.

Measurement

1. Principle

Alkalinity is measured as carbonate, bicarbonate, and hydroxide ions which resulted during dissociation or hydrolysis of solutes, which, in turn, react with standard acid, added to the sample.

2. Apparatus

a. Titration assembly — comprise the following:
    i. Volumetric flasks — from 100-ml to 1L capacity
    ii. Volumetric pipettes — graduated
    iii. Burette — 25- or 50-ml capacity, fixed on a clamped stand.
    iv. Erlenmeyer flasks — 250-ml capacity
b. Oven — set at 250°C
c. Desiccating cabinet
d. Magnetic stirrer

3. Reagents

a. Sodium carbonate solution; Na_2CO_3, 0.05N.
   i. Dry an aliquot quantity of Na_2CO_3 in an oven at 250°C for 4 h. Keep in a desiccating cabinet to bring it to room temperature.
   ii. Weigh accurately 2.5g and transfer to a 1L volumetric flask. Dissolve in DDW.
   iii. Dilute it up to the mark with DDW.

Note: Prepare fresh solution of 0.05N Na_2CO_3 weekly.

b. Stock solution of H_2SO_4 or HCl, 0.1N: Refer to Section 5.3.5 and Table 3.4.

c. Standard solution of H_2SO_4 or HCl, 0.02N:

Measure 100 ml of 0.1N stock acid solution in a 500-ml volumetric flask. Bring the level to the mark with DDW.

$$1 \text{ ml } 0.02 \text{ acid solution} \equiv 1.0 \text{ mg } CaCO_3$$

d. Phenolphthalein indicator: Weigh 0.5-g phenolphthalein disodium salt and transfer to a 100-ml volumetric flask. Dissolve in 1:1 ethanol and DDW solution. Dilute to the mark with DDW.

e. Mixed indicator: Weigh 0.1-g bromocresol green sodium salt and 0.02-mg methyl red. Transfer this mixture to a 100-ml volumetric flask. Dissolve in 95% ethyl alcohol or DDW. The final volume should be 100 ml.

f. Methyl orange indicator: Weigh 0.05 g powdered methyl orange and transfer to a 100-ml volumetric flask. Dissolve in DDW and dilute to 100 ml. If sodium salt of methyl orange is available, dissolve 0.05-g sodium salt in 100-ml DDW. Add 1.5 ml of 0.1 M HCl (Table 3.4) and filter if any turbidity appears.

4. Standardization of H_2SO_4 titrant, 0.02 N

Always prepare the sample and blank in duplicate.

a. Measure accurately 10 ml of 0.05N Na_2CO_3 into a 250-ml conical flask and add 90 ml DDW.

b. Add 0.1 ml methyl orange indicator.

c. Titrate with 0.02N H_2SO_4. Approximately 24 ml of the titrant will be used to a 4.5 pH end point (Table 7.1).

d. Prepare a blank using 100-ml DDW instead of standard solution and repeat the same procedure of titration as described for standard. Record the volume of titrant used in both cases.

e. Calculate the normality of titrant as follows:

$$\text{Normality of titrant} = \frac{A \times 10.0}{53.0 \times (C - B)} \approx 0.02 \text{ N}$$

TABLE 7.1
Alkalinity Color Indicators and pH End Points Relationship

Indicator	pH at end point	Color change	Volume of indicator per 100 ml sample
Phenolphthalein	8.3	Pink to colorless	0.1
Mixed	5.0	Greenish blue to light blue with lavender gray	0.15
	4.8	Greenish blue to light pink gray with bluish cast	
	4.6	Greenish blue to light pink	
Methyl orange	4.6	Yellow to orange	0.5
	4.0	Yellow to pink	

Adapted from Adams, V. D., *Water and Wastewater Laboratory Manual*, Lewis Publishers, Inc., 1990, with permission.

where A = weight of Na_2CO_3 in g used to prepare 1L of 0.05 N solution
B = volume of acid used for the blank, ml
C = volume of acid used for the standard, ml

5. Procedure

a. Take 100 ml of unfiltered sample or an aliquot diluted to 100 ml into a 250-ml Erlenmeyer flask.

Note: Dilute the sample if alkalinity is high, i.e., if it requires more than 25 ml of titrant.

b. Keep the beaker on a magnetic stirrer. Insert bar magnet and pH meter electrodes in the beaker. Turn on the stirrer. Stir the sample with the magnetic stirrer so that bar magnet does not touch the electrode.
c. Record initial pH and temperature.
d. If initial pH is more than 8.3, titrate to 8.3 with 0.02N H_2SO_4 using phenolphthalein indicator, till the pink color disappears.
e. Record the volume of the titrant used when the red color disappears. This is phenolphthalein end point (P).
f. Continue titration by adding 2–3 drops of mixed indicator until the color of indicator changes from green to red. This is the end point.
g. Record the volume of the titrant. This is the total alkalinity (T).
h. Prepare a blank with 100 ml DDW and run the titration by following the same procedure as described for the sample. Blank should be run in duplicate.

6. Calculation

$$\text{Alkalinity, mg/L as } CaCO_3 = \frac{(A - B) \times N \times 50000}{\text{volume of sample, ml}}$$

where A = volume of 0.02 N H_2SO_4 used for sample, ml
B = volume of 0.02 N H_2SO_4 used for Blank, ml
N = normality of acid

7. Relationship of alkalinity

The results obtained from the phenolphthalein (P) and total alkalinity (T) are related and sitochiometrically classify the three forms of alkalinity present in water and wastewater, i.e., alkalinity due to hydroxide, carbonate and bicarbonate ions. The complete relationship scheme is given in Table 7.2.

TABLE 7.2
Alkalinity Relationship

Result of titration	OH^- alkalinity as $CaCO_3$	CO_3^{2-} alkalinity as $CaCO_3$	HCO_3^- alkalinity as $CaCO_3$
P = 0	0	0	T
P < $^1/_2$ T	0	2P	T - 2P
P = $^1/_2$ T	0	2P	0
P > $^1/_2$ T	2P — T	2(P - T)	0
P = T	T	0	0

Where P - phenolphthalein alkalinity
T - total alkalinity

According to this scheme:

a. CO_3^{2-} alkalinity is present when P is not zero but is less than or equal to half T.
b. OH^- alkalinity is present if P is more than half T.
c. HCO_3^- ions are present if P is equal to zero or less than half T.

Adapted from APHA, *Standard Methods for the Examination of Water and Wastewater*, 17th Ed., APHA, 1989, with permission.

7.3 Acidity — Titrimetric Method

Acidity of water or wastewater is *quantitative* representation of its ability to neutralize strong bases.

Source

Acidity in water and wastewater can be imparted by following three sources:

1. Dissolution of CO_2

Carbon dioxide is a common constituent of all natural waters. It can access surface waters by the following means:

a. Absorption from the environment due to the difference in partial pressure of CO_2 in water and in the atmosphere.

b. It is one of the end products produced during the biological decomposition of organic matter available in waters — especially in wastewater — under both aerobic and anaerobic conditions.

c. Sometimes ground water contains more than 50 mg/L CO_2. This water, when percolated through soils that do not contain enough $MgCO_3$ or $CaCO_3$ to neutralize CO_2 in the form of bicarbonates, contributes acidity.

$$CO_2 + CaCO_3 + H_2O \rightarrow Ca(HCO_3)_2 \tag{7.13}$$

Dissolution of CO_2 in water gives rise to a common buffer system H_2CO_3 — HCO_3^- (carbonate – bicarbonate) which is responsible for the acid–base balance in water and wastewater.

2. Acidity due to mineral acids

a. When the effluent of some industries involved in metallurgical processes and in the production of synthetic organic materials is discharged to water streams, mineral acidity can be imparted to water.

b. The discharge of mines containing sulfur, sulfides, or iron pyrites may produce H_2SO_4 or salts of sulfuric acid by the action of sulfur oxidizing bacteria under aerobic conditions. This process imparts acidity to water. Two examples are cited in Equations 7.14 and 7.15.

$$2\,S + 3\,O_2 + H_2O \xrightarrow[\text{bacteria}]{\text{sulfur oxidizing}} 2\,H_2SO_4 \tag{7.14}$$

$$2\,Fe\,S_2 + 7\,O_2 + 2\,H_2O \xrightarrow{\text{bacteria}} 2\,FeSO_4 + 2\,H_2SO_4 \tag{7.15}$$

3. Heavy metal Salts

Salts of some heavy metals like Fe^{3+}, Al^{3+} on hydrolysis yield proton (H^+) which causes acidity in water.

$$Al\,Cl_3 + 3\,H_2O \leftrightarrow Al\,(OH)_3 + H^+ + Cl^- \tag{7.16}$$

Significance

1. Corrosion

Water containing high acidity is of great concern because of its corrosive nature, which can cause the destruction of water mains, sewers, and other related equipment.

2. Wastewater treatment

Acidity in water influences the biological processes of the treatment of wastewater. If an effluent with high acidity is discharged into wastewater, it lowers the pH of the biological system from its normal range, which lies between 6.0 and 8.5. High acidity reduces the activity of aerobic bacteria carrying out the oxidation of organic wastes. This affects the treatment of wastewater and the quality of treated effluent produced. Hence, for appropriate biological oxidation of organic wastes, the adjustment of pH is required. This may increase the cost of treatment process.

3. Water treatment

Acidity has influence on chemical treatment of water also. Excess of CO_2 interferes with the water-softening process, where the lime or lime-soda ash method is used.

Measurement

1. Principle

Acidity in water and wastewater is the measure of acid radicals (produced by dissociation of both strong and weak acids) present in the samples. This is detected by neutralization with strong base. Phenolphthalein is used as an indicator to detect the end point of neutralization. Acidity is expressed as mg $CaCO_3$ / L.

2. Apparatus

All apparatus as described in Section 7.2.

3. Reagents

All reagents should be prepared in CO_2-free water.

a. Solvent, CO_2-free water: Boil fresh DDW for 15 min. and cool to room temperature.
b. Potassium hydrogen phthalate, (KHP) $KH(C_8H_4O_4)$, solution; 0.05 N:
   i. Dry an adequate quantity of analytical grade KHP in an oven at 110°C for 2 h. Cool to room temperature by keeping in a desiccating cabinet.
   ii. Weigh accurately 10.21 g KHP and transfer to a 1-L volumetric flask. Dissolve in CO_2-free DDW. Dilute up to the mark with CO_2-free DDW.
   iii. Store this solution in a tightly closed glass reagent bottle.

Note: This solution is stable for about a month. If the formation of mold or turbidity appears, discard the solution and prepare fresh standard KHP solution.

c. Stock sodium hydroxide solution; 0.1 N: Refer to Section 3.5.3 and Table 3.5.

$$1 \text{ ml of NaOH solution} \equiv 5.0 \text{ mg } CaCO_3 \text{ acidity}$$

d. Standard sodium hydroxide solution; 0.02 N:

Measure 200 ml of 0.1 N NaOH solution in a 1-L volumetric flask and dilute up to the calibration mark with CO_2-free DDW.

Note: Store the reagent in a polyethylene bottle with a tight screw cap. The normality of NaOH will change with time if it is not protected from absorption of atmospheric CO_2. Take the following precautions:

(1) Attach a tube filled with soda lime to the reagent bottle containing 0.02 N NaOH solution. Soda lime will absorb atmospheric CO_2.

(2) Always dispense the solution by an automatic dispenser in order to avoid opening the bottle.

e. Stock sulfuric acid, H_2SO_4, 0.10 N: Refer to Section 7.2.
f. Standard sulfuric acid, 0.02 N: Prepare and standardize this solution in the same manner as described in Section 7.2.
g. Hydrogen peroxide, H_2O_2, 30%: Commercially available solution. Store in a refrigerator, as it is a strong oxidizing agent.
h. Sodium thiosulfate solution; $Na_2S_2O_3$, 0.1 N: Weigh accurately 12.5 g $Na_2S_2O_3 \cdot 5\ H_2O$ and transfer to a 500-ml volumetric flask. Dissolve in CO_2-free DDW. Dilute up to the calibration mark.
i. Phenolphthalein indicator: Refer to Section 7.2.

4. Standardization

a. Standardization of NaOH Titrant; 0.02 N:
   i. Measure 10 ml of 0.05 N KHP in a 250-ml Erlenmeyer flask and add 40 ml CO_2-free DDW. Prepare in duplicate.
   ii. Keep calibrated pH electrodes in the beaker. Titrate KHP solution with NaOH titrant until pH 8.7 is achieved using phenolphthalein as indicator.

Note: 10 ml KHP solution will require about 25 ml of NaOH titrant.

   iii. Prepare a reagent blank using 50-ml DDW instead of KHP solution and follow the same procedure of titration as described for KHP solution.

iv. Calculate the actual normality of the NaOH titrant by using following relation:

$$\text{normality} = \frac{A \times 10}{204.2 \times B - C} \approx 0.2 \text{ N}$$

where A = weight of KHP in g used to prepare 1L of 0.05 N solution
B = volume of NaOH titrant used for KHP solution, ml
C = volume of NaOH titrant used for blank, ml

Note: *(1) For every estimation, standardize 0.02 N NaOH solution as it will degrade on storing. (2) The average blank value, C, is used again in the sample calculations.*

b. Standardization of H_2SO_4; 0.02 N: Follow the procedure described in Section 7.2.

5. Procedure

a. Sample Preparation

i. If sample contains residual chlorine, neutralize it with 0.1 N $Na_2S_2O_3$ solution according to the following relation:

$$1 \text{ mg residual chlorine} \equiv 1 \text{ ml } 0.1 \text{ N } Na_2S_2O_3 \text{ solution}$$

ii. If the sample is suspected to contain industrial waste rich in hydrolyzable metal ions or reduced forms of polyvalent heavy metals, add 1–2 ml 30% H_2O_2 per 100 ml of the sample. Boil the sample for 2–5 min. Cool to room temperature.

b. Estimation

i. Take 50 ml of sample (unfiltered) or an aliquot diluted to 50 ml with CO_2-free DDW in a 125-ml beaker.

ii. Keep the beaker on a magnetic stirrer. Insert a calibrated pH meter probe and magnetic stirring bar.

iii. Record the pH of the sample.

iv. If the pH of sample is more than 4.0, add 0.02 N H_2SO_4 in small increments to lower the pH to ≥ 4.0. Record the volume of acid used.

v. Remove the pH probe.

vi. Titrate with 0.02 N NaOH titrant to pH 8.3 by adding the titrant in small increments to detect the end point accurately using phenolphthalein indicator.

vii. Record the volume of alkali added.

6. Calculations

$$\text{Acidity as } CaCO_3\text{, mg/L} = \frac{[(A - B) \times N - (D \times E)] \times 50000}{\text{Volume of sample, ml}}$$

where A = volume of 0.02 N NaOH titrant used for sample, ml
B = volume of 0.02 N NaOH titrant used for blank, ml
N = normality of standard NaOH titrant
D = volume of H_2SO_4 used, ml (Note – the term may be zero if pH of sample is <4.0)
E = actual Normality of standard H_2SO_4

7.4 Hardness — EDTA Titrimetric method

Hardness of water is defined as the presence of significant concentration of salts of metallic cations mainly Ca^{2+} and Mg^{2+} ions dissolved in water. Under supersaturated conditions, these cations react with anions to form insoluble solid precipitate. Hardness is classified into two types.

1. Carbonate hardness

This type of hardness is due to the presence of calcium and magnesium carbonates and bicarbonates in water. It is expressed in terms of $CaCO_3$ concentration in mg/L. This is also known as temporary hardness because it is highly sensitive to heat and precipitates out readily on boiling.

$$Ca(HCO_3)_2 + Heat \rightarrow CaCO_3 + CO_2 + H_2O \quad (7.17)$$

$$Mg\ (HCO_3) + Heat \rightarrow Mg\ (OH)_2 + 2\ CO_2 \quad (7.18)$$

2. Noncarbonate hardness

This type of hardness in water occurs due to dissolution of salts of calcium other than carbonates and bicarbonates, such as calcium sulfate ($CaSO_4$) or calcium fluoride (CaF_2). This hardness is referred as Permanent Hardness because it cannot be removed by boiling.

Sources

Hardness in water and wastewater is mainly due to the presence of multivalent cations like Ca^{2+}, Mg^{2+}, Fe^{2+} and Mn^{2+} (reduced forms), St^{2+}, and Al^{3+}. Out of these, Ca^{2+} and Mg^{2+} ions are most abundant, while others occur in insignificant amounts. Therefore, hardness in water is largely due to the availability of Ca^{2+} and Mg^{2+} ions. Thus, hardness can be represented by the sum of both cations.

Significance

Hardness of water is an essential parameter in assessing the suitability of water for domestic and industrial uses for the following reasons.

1. Lather formation

Hard water does not produce lather or foam with sodium soaps, thus resulting in economic loss to water consumers. Sodium soaps precipitate with the metallic cation (M) that causes hardness in waters, thus losing their surfactant properties.

$$2NaCO_2C_{17}H_{33} + M^{2+} \rightarrow M^{2+}(COOC_{17}H_{33})_2 + 2Na^+ \quad (7.19)$$

This precipitate adheres to tubs, sinks, and dishwashers and causes stains on clothes, dishes, and other items. When hard water is used for bathing, the precipitate may remain in the pores and the skin may feel rough and uncomfortable.

2. Efficiency of boilers and heating systems

Hard water reduces the efficiency of boilers and heating systems in various industries due to the scale that is deposited on the equipment. This causes equipment breakdown and considerable economic loss.

3. Public health

Magnesium hardness, particularly associated with the sulfate ion, has a laxative effect on persons unaccustomed to it. Calcium hardness does not cause any public health problem. In fact, hard water is apparently beneficial to the human cardiovascular system.

Measurement

1. Principle

Ca^{2+} and Mg^{2+} ions, the major causes of hardness in water, form a soluble chelated complex with EDTA (ethylene diamine tetra acetic acid). Eriochrome black T, a dye, is used as an indicator. When the indicator is added to hard water at pH of 10.0 ± 0.1, the solution becomes wine red, due to the formation of a complex between metal ions and indicator.

$$\underset{}{M^{2-} + I(\text{Indicator}) \rightarrow \underset{(\text{Wine red complex})}{M\,I^{2-}}} \quad (7.20)$$

When EDTA is added to the solution, color changes from wine red to blue at the end point. This is because EDTA breaks up the wine-red complex (MI^{2-}) and chelate with divalent cations, releasing the free indicator molecules The complex formed between divalent cations and EDTA is blue in color between pH 7 and 11. Therefore, the solution turns blue at the end point, as shown in Equation 7.21.

$$M\,I^{2} + EDTA \rightarrow [M.EDTA]\ \text{complex} + I^{2-} \quad \text{(blue)} \tag{7.21}$$

To minimize the chances of $CaCO_3$ precipitation, the titration should not take more than 5 min at room temperature because the indicator works best at this temperature.

Note: The titration at room temperature must be finished within 15 min to prevent the loss of $CaCO_3$ by precipitation.

2. Apparatus

All apparatus as described in Section 7.2.

3. Reagents

a. Buffer solution

   i. Solution (a) - Weigh 2.358 g of EDTA disodium salt and 1.56 g of $MgSO_4.7H_2O$ (if Mg salt of EDTA is available, use 2.5 g) and transfer to a 100 ml volumetric flask. Dissolve the mixture in DDW and bring the volume up to the mark with DDW.

   ii. Solution (b) - Weigh 33.8-g ammonium chloride (NH_4Cl) and transfer to a 500-ml volumetric flask. Dissolve in 284 ml of concentrated ammonium hydroxide (NH_4OH).

Upon complete dissolution, add solution (a) to solution (b) and mix thoroughly. Finally, bring the volume up to 500 ml with DDW. The pH of the buffer solution should be 10.0 ±0.1.

Note: (1) Store the buffer solution in plastic or Pyrex® container.

(2) Prepare a fresh solution every two weeks.

b. Interference

Some polyvalent cations like Al, Ba, Cd, Co, Cu, Fe, Pb, Mn, Ni, Sr, and Zn may interfere with the titration with EDTA by developing faded or indistinct color at the end point. Chelating agents listed below can combine with the interfering ions and help to get a clear, sharp change in color at the end point. Thus, they can remove cations interfering with the analysis.[10]

   i. Chelating agent I, sodium cyanide, NaCN: adjust the pH of the sample to 6 with buffer or higher with 0.1N NaOH. Add 250 mg of NaCN (powder) to the sample. Add enough buffer to adjust the pH to 10.0 (0.1).

Note: NaCN is highly toxic, so proper precaution must be taken during the use and disposal of the solution containing NaCN. Never transfer this solution with pipette. Always transfer with the help of a burette.

ii. Chelating agent II, sodium sulfide, Na_2S. 5 H_2O: dissolve 3.7 g Na_2S. 5 H_2O in 100 ml DDW.

Note: *Store this reagent in a tightly stoppered bottle because it is unstable and readily oxidizes when exposed to air. Add 1 ml of reagent to the sample before titration.*

iii. Chelating agent III, Mg salt of 1,2-Cyclohexanediamine tetra acetic acid, CDTA: add 0.25 g of Mg CDTA to 100 ml sample and dissolve completely before adding buffer solution. This chelating agent is preferred because of its nontoxic nature.

c. Indicator — Eriochrome Black T

Weigh 0.5 g Eriochrome Black T dye and 4.5 g hydroxylamine hydrochloride. Transfer this mixture to a 100-ml volumetric flask. Dissolve it in 95% ethyl or isopropyl alcohol. Make up the final volume to 100 ml with alcohol.

d. Standard EDTA Titrant, 0.01M

Accurately weigh 3.723-g EDTA disodium salt and transfer to a 1-L volumetric flask. Dissolve in DDW. Bring the volume up to the mark with DDW. Store in a polythene or glass bottle.

e. Hydrochloric acid, HCl, 50% (v/v)

Measure 50 ml DDW in a 100-ml volumetric flask and add 50 ml concentrated HCl in the flask with stirring. Cool the solution.

f. Ammonium hydroxide, NH_4OH, 24% (v/v)

Measure 240 ml concentrated NH_4OH in a 1-L volumetric flask and dilute to 1 L with DDW

g. Standard Calcium solution, $CaCO_3$ anhydrous

i. Weigh accurately 1.0 g anhydrous $CaCO_3$ and transfer to a 500-ml Erlenmeyer flask.

ii. Place a funnel in the flask and add 50% (v/v) HCl in small increments to dissolve $CaCO_3$ completely.

iii. Add about 250 ml DDW and boil the solution for few minutes to expel CO_2 from the system. Cool to room temperature.

iv. Adjust the pH of the solution nearly to 5 by adding either 50% (v/v) HCl or 24% (v/v) NH_4OH solution according to the requirement.

v. Transfer this solution to a 1L volumetric flask. Rinse the Erlenmeyer flask several times with DDW and add the rinsing liquid to the volumetric flask. Make up the final volume to 1L with DDW.

$$1 \text{ ml standard solution} = 1 \text{ mg } CaCO_3$$

4. *Standardization*

Prepare standard calcium solutions containing 50, 100, and 150 mg $CaCO_3$/L concentration and standardize by following procedure:

a. Measure into three different 100-ml volumetric flasks 5, 10, and 15 ml of standard $CaCO_3$ solution.
b. Dilute the solution in each flask up to the mark with DDW.
c. Add about 5-ml buffer and examine the pH. It must be 10 ± 0.1.
d. Add 0.5 ml of indicator solution in each flask.
e. Titrate with standard EDTA solution. Standard $CaCO_3$ solutions will consume approximately 5, 10, and 15 ml of the EDTA titrant, respectively.

$$1.0 \text{ ml } 0.01 \text{ M EDTA} \equiv 1.0 \text{ mg } CaCO_3$$

f. Similarly prepare a blank by taking 100 ml DDW instead of standard solution and follow the same procedure of titration. Blank will consume less than 0.5 ml of the titrant.

5. *Procedure*

a. Take 100-ml sample or aliquot quantity of sample diluted to 100 ml in a 250 ml Erlenmeyer flask. Add about 5 ml buffer to acquire a pH level of 10.0 ± 0.1.

Note: *If the sample has hardness more than the range, it must be diluted so that the volume of titrant consumed falls within the range of 5 to 15 ml.*

b. Add 0.5-ml indicator. This will give a wine-red color to the solution.
c. Titrate slowly with EDTA, stirring constantly, until the reddish tinge disappears. Add the last few drops at an interval of 3–5 s. At the end point, the solution turns blue.
d. Record the volume of the titrant consumed.

6. *Calculations*

$$\text{Total hardness as } CaCO_3(\text{mg/L}) = \frac{(A - B) \times D \times 1000}{\text{volume of sample, ml}}$$

where A = volume of titrant consumed for sample, ml
B = volume of titrant consumed for blank, ml
C = volume of titrant consumed for standard, ml
D = mg $CaCO_3$ equivalent to 1.00 ml EDTA titrant

$$\therefore D = \frac{\text{mg } CaCO_3}{(C - B)} \approx 1$$

7.5 Nonmetallic constituents

This section describes the measurement of nonmetallic constituents of water and wastewater. These measurements are conducted for the assessment and control of potable and receiving water quality. They are also useful in determining the process efficiency of wastewater treatment plants.

7.5.1 Boron — Carmine Colorimetric Method — Detectable Range: 100 to 1000 μg/L Boron

Source

Boron may occur in natural waters, but significant quantities are received through discharge of industrial wastes. Boron pollution occurs mainly from the discharge of industries using or manufacturing synthetic boranes such as detergents, steel, glass, and pesticides. Sometimes marine water contains about 5 mg B/L.[8]

Significance

Toxicity. Boron is toxic for many organisms in concentrations as low as 1 mg\L, while other organisms tolerate levels up to 15-mg/L. It is an essential nutrient for the growth of plants if present up to 0.5 mg\L. If the level in irrigation water exceeds this value, it causes the destruction of sensitive plants such as apples, citrus fruits, and nuts etc. Generally, water containing 2-mg\L concentration is considered unsatisfactory for irrigation. The concentration of boron less than 0.1 mg\L is considered safe for human consumption.[8] If it is ingested in a large concentration, it can affect the central nervous system. Prolonged ingestion can result in a clinical syndrome known as borism.[11]

Measurement

1. Principle

Boron available in water or wastewater sample reacts with carmine under acidic conditions. The bright red color of the reagent changes to a bluish red or deep blue during the reaction. The intensity of blue color is proportional to the concentration of boron available in the sample.

2. Apparatus

a. Spectrophotometer — Set at 605 nm wavelength
b. Volumetric flasks — 100 ml and 1 L capacity
c. Beakers — Polyethylene, 100 and 150 ml capacity

3. Reagents

a. Hydrochloric acid: concentrated HCl.

b. Sulfuric acid: concentrated H_2SO_4

c. Carmine reagent: Weigh 0.046 g Carmine (Alum Lake) 40 and transfer to a 100-ml volumetric flask. Dissolve in concentrated H_2SO_4 and make up the final volume to 100 ml. Keep overnight for complete dissolution. After preparation, store in a refrigerator.

d. Stock boric acid solution, H_3BO_3: weigh 0.5722 g anhydrous H_3BO_3 and transfer to a 1-L volumetric flask. Dissolve in DDW. Make up the final volume to 1L with DDW.

$$1.0 \text{ ml stock solution} \equiv 100 \ \mu\text{g boron}$$

e. Standard boric acid solution: measure 1.0 ml of stock boric acid solution in a 100-ml volumetric flask and dilute up to the mark with DDW.

$$1.0 \text{ ml standard solution} \equiv 1.0 \ \mu\text{g boron}$$

Note: *(1) Follow rigorous safety rules, as concentrated acids are used in the estimation.*

(2) Store all the reagents in polyethylene or other plastic boron-free containers.

(3) Use polyethylene or boron-free glassware for estimation.

4. Standardization

Prepare calibration solutions of boron containing 100, 250, 500, 750, and 1000 µg B/L concentration from standard boric acid solution and standardize as follows:

a. Into four different 100-ml volumetric flasks, measure 10, 25, 50, and 75 ml standard boric acid solution.

b. Dilute solution in each flask up to the mark with DDW.

c. For 1000 µg B/L concentration, use the standard solution directly without dilution.

d. Measure 2.0 ml of each above-mentioned standard into five different 100-ml plastic beakers. Place each beaker in a hood.

e. Add carefully 0.10 ml concentrated HCl and mix well.

f. Then introduce 10 ml concentrated H_2SO_4 to each standard in small increments carefully. Mix thoroughly after addition of each increment of acid.

g. Cool each standard solution to room temperature.

h. Add 10.0 ml carmine solution to each standard after attaining room temperature and mix thoroughly. Wait for 2 h for complete reaction.

i. Prepare a blank with 2.0-ml DDW instead of standard solution and set zero in the spectrophotometer.

j. Mix all the standards well before pouring into the spectrophotometric cell.

k. Measure the absorbance of standards against blank at 605 nm. Prepare a calibration curve of absorbance against concentration of boron in mg/L.

5. Procedure

a. Rinse 150-ml plastic beakers the day before with DDW and drain.

b. Allow the carmine reagent to come to room temperature.

c. Measure the boron content in the samples by following the same procedure as described for the standards.

6. Calculation

Calculate the concentration of boron in a sample from the calibration curve or from the direct readout of the instrument.

7. Important instructions

a. Bring the entire sample, standards, and carmine reagent to room temperature before addition of reagent to the samples.

b. Rinse the spectrophotometric cells with sample before pouring the sample into it.

c. Do not rinse the cell with DDW in between the sample measurements. Rinsing should be done with the sample only.

d. Avoid bubble entrapment in the sample while making photometric readings.

e. Examine the calibration daily, as carmine deteriorates readily.

7.5.2 Chloride — Mercuric Nitrate Titration Method

Source

1. Natural waters

Chlorides are the common constituents of all natural waters. This may be due to leaching of chloride containing rocks and soils, which come in contact with water in areas adjacent to the ocean.

Marine water is also highly rich in chloride content. Hence, ground waters in coastal areas show significant high level of chlorides.

2. Sewage

Sewage produced after domestic use of water is also a rich source of chloride because of the significant chloride concentration in human wastes. Human excreta, mainly urine, is rich in chlorides. Normally, 5–8g of chlorides per day per person access sewage.[12] Thus, domestic wastewater adds a considerable amount of chlorides to the receiving water streams.

A high level of chlorides in a water stream indicates that the water body is being used as a human-waste disposal site.

3. Industrial wastes

Effluents generated from oil-field operations or papermaking, galvanizing, and water-conditioning industries transfer significant amounts of chlorides to a water stream. In addition, wastewater from agricultural sites discharged to surface waters release a significant amount of chlorides into the water.

Significance

1. Toxicity and palatability

Chlorides are not harmful or toxic to humans even at concentrations as high as 2,000 mg/L. Therefore, they do not present a major pollution threat. The higher concentrations of chloride impart an objectionable salty taste to water, which affects its palatability, making it unacceptable for public consumption. The permitted level is 250 mg/L in potable water. A concentration less than 25 mg/L in drinking water is considered to be an ideal condition.[8]

2. Irrigation

Chloride ion concentration is an important factor to be considered if treated effluent is used for irrigation. For this purpose the concentration should not be more than 140 and 100 mg/L, if plants are irrigated by a surface irrigation method, or by a sprinkler irrigation method.[13] High chloride concentration disturbs the osmotic balance between the plants and the soil, which affects the growth of plants.

3. Cooling water

The use of water containing high chloride content as cooling water in boilers and other heating systems is restricted in industries. It affects the metallic pipes and related equipment.

Measurement

1. Principle

a. The level of chlorides in water or wastewater sample (pH is adjusted to 2.5) is determined by the titration of the sample with mercuric nitrate in the presence of an indicator. Under these conditions, Hg^{2+} ion combines with Cl^- ions to form mercuric chloride ($HgCl_2$) complex.

$$Hg^{2+} + 2Cl^- \leftrightarrow HgCl_2 \ (K = 2.6 \times 10^{-15}) \qquad (7.22)$$

b. Mercuric chloride complex is soluble; therefore, diphenyl carbazone is used as an indicator to obtain a sharp end point. The indicator combines with excess Hg^{2+} ions and produces a distinct purple color at pH 2.5.

c. The sharpness of end point is further improved by adding xylene cyanol FF. This indicator is blue-green at pH 2.5. At end point, its color changes from blue-green to blue and then to purple. Hence a more distinct color change is obtained at end point.

d. Nitric acid is added to the indicator to reduce the pH of samples, standards, and blank to 2.5.

2. Apparatus

a. Volumetric flasks — 100 ml and 1 L capacity

b. Erlenmeyer flasks — 250 ml capacity

c. Oven — set at 120°C

d. Desiccating cabinet

3. Reagents

a. Nitric acid: Concentrated

b. Sodium bicarbonate, $NaHCO_3$: powder

c. Acidified Indicator reagent:

   i. Weigh 0.25-g S-diphenyl carbazone and transfer to a 100-ml volumetric flask. Dissolve in small quantity of 95% ethyl alcohol.

   ii. Add 4.0-ml concentrate HNO_3 to the flask.

   iii. Weigh 0.03-g xylene cyanol FF and dissolve in the above solution.

   iv. Make up the final volume to 100 ml with 95% ethyl alcohol and store in a dark reagent bottle in the refrigerator.

d. Mercuric nitrate titrant, 0.0141 N: keep an adequate quantity of mercuric nitrate in a desiccating cabinet for 1 h. Weigh 5.75 g $Hg(NO_3)_2$ or 6.25 g $Hg(NO_3)_2 \cdot H_2O$ and transfer to a 250-ml volumetric flask. Dissolve in 250 ml DDW containing 1.25-ml concentrate HNO_3. Store this solution in a dark reagent bottle.

$$1.0 \text{ ml } 0.0141 \text{ N mercuric nitrate titrant} \equiv 0.5 \text{ mg } Cl^-$$

Note: *Always store the chemical in the dark in a desiccating cabinet. The crystals, which absorb moisture, must not be used for making titrant solution.*

e. Standard sodium chloride, 0.0141 N: Dry an adequate quantity of NaCl in an oven set at 120°C for 2 h. Cool to room temperature in a desiccating cabinet. Accurately weigh 1.6482 g NaCl and transfer to a 1-L volumetric flask. Dissolve in DDW. Dilute up to the mark with DDW.

$$1.00 \text{ ml standard NaCL solution} \equiv 1.0 \text{ mg } Cl^-$$

4. *Standardization*

Prepare calibration standards of NaCl containing 25-, 50-, and 100-mg/L chloride concentration and standardize as follows:

a. Into three different 250-ml Erlenmeyer flasks, measure 2.5, 5, and 10 ml of the standard NaCl solution.

b. Add 10 mg $NaHCO_3$ into each Erlenmeyer flask. Bring the final volume to 100 ml in each flask with DDW.

c. Take another 250-ml Erlenmeyer flask, measure 100 ml DDW and add 10 mg $NaHCO_3$ into it. This will serve as a blank.

d. Add 1.0 ml of the acidified indicator reagent in all the flasks and mix well. The pH of the solution should be 2.5 ± 0.1.

e. Titrate with mercuric nitrate titrant till a purple color is obtained. This is the end point.

Note: *Preserve these standards to compare the intensity of purple color developed in the blank and samples.*

5. *Procedure*

a. Measure a 100-ml water or wastewater sample into a 250-ml Erlenmeyer flask. If the chloride content is expected to be high, dilute the sample with DDW so that it will consume not more than 10-ml titrant.

b. Add 1 ml of acidified indicator reagent. The pH should be 2.5 ± 0.1. At this pH, indicator gives a blue-green color to the sample.

c. Titrate with mercuric nitrate titrant following the same procedure as described under Standardization.

6. *Calculations*

$$\text{mg/L Cl}^- = \frac{(A - B) \times N \times 35.5 \times 1000}{\text{volume of sample, ml}}$$

where A = volume of titrant used for the sample, ml
B = volume of titrant used for the blank, ml
N = normality of mercuric nitrate titrant

7. *Important instructions*

a. The acidified indicator should contain enough HNO_3, to completely neutralize the total alkalinity and maintain pH at 2.5 ± 0.1 in the sample. If the sample is alkaline, either dilute or neutralize.

b. $NaHCO_3$ is used in the blank and standards for maintenance of proper pH upon addition of acidified indicator.

c. Mercury is a hazardous material; hence, wastes containing it should be disposed of with care.

7.5.3. Residual and Total Chlorine — DPD Colorimetric Method

Chlorine can be present as free chlorine or in the combined state. Both forms can exist in the same water sample and be determined as total chlorine. The fraction of chlorine available as hypochlorous acid or hypochlorite ion represents the free or residual chlorine content. The rest, which exists in the form of chloramines or other chloro derivatives, falls under combined chlorine category.

Source

Free or residual chlorine available in public water supplies and wastewater effluents is due to the chlorination of water, which is essential for the control of pathogenic organisms.

Chemistry of the chlorination process

Chlorine is added to water in the form of free liquid chlorine or hypochlorites [NaOCl or Ca $(OCl)_2$] for the chlorination of water.

In water, liquid chlorine combines to form hypochlorous and hydrochloric acid.

$$Cl_2 + H_2O \leftrightarrow HOCl + HCl \quad (7.23)$$

Further, the hypochlorous acid, HOCl, is a weak acid. It ionizes or dissociates into H^+ and hypochlorite ion at pH less than 6.0.

$$HOCl \leftrightarrow H^+ + OCl^- \quad (7.24)$$

The quantity of HOCl and OCl^- present in water is known as free available chlorine, which carries out disinfection, i.e., the control of pathogenic bacteria. Reactions 7.23 and 7.24 are pH dependent. As a general rule, chlorine and its products are most effective disinfectants at low pH. At pH values less than 3.0, very little Cl_2 exists in molecular form.

The reactive species OCl^- can also be produced by dissociation of hypochlorite salts like calcium and sodium hypochlorite as shown in the Equations 7.25 and 7.26.

$$Ca\,(OCl)_2 \rightarrow Ca^{2+} + 2\,OCl^- \quad (7.25)$$

$$NaOCl \rightarrow Na^+ + OCl^- \quad (7.26)$$

The OCl^- ions establish equilibrium with H^+ ions, as shown in Equation 7.24. In both cases of disinfection, equilibrium establishes between OCl^- and H^+ ions irrespective of the disinfecting agent, whether liquid chlorine or hypochlorites. The major difference lies in the pH of the system, which decides the existence of relative quantities of OCl^- and HOCl species at equilibrium. This accordingly represents the efficiency of the disinfection system.

Reactions of chlorine with different pollutants in water

Chlorine and hypochlorous acid react with a wide range of pollutants present in water and wastewater according to the reactions described below:

1. Ammonia

The treated wastewater effluent contains significant amounts of nitrogen in the form of NH_3 or nitrate if the treatment plant operation is designed to achieve nitrification. Hypochlorous acid is a strong oxidizing agent, hence oxidizes ammonia into three types of chloramines as mentioned in the successive reactions.

$$NH_3 + HOCl \rightarrow \underset{\text{(monochloramine)}}{NH_2Cl} + H_2O \tag{7.27}$$

$$NH_2Cl + HOCl \rightarrow \underset{\text{(dichloramine)}}{NHCl_2} + H_2O \tag{7.28}$$

$$NHCl_2 + HOCl \rightarrow \underset{\text{(trichloramine)}}{NCl_3} + H_2O \tag{7.29}$$

The chlorine available in these compounds is known as combined available chlorine. These reactions depend on pH, temperature, contact time, and the concentration of liquid chlorine or hypochlorites taken in proportion to the concentration of ammonia.

The mono- and dichloramines have significant disinfecting power, but it is inferior to that of hypochlorite ion and hypochlorous acid, hence it is important to assess the exact requirement of chlorine or hypochlorites for disinfection. In the presence of NH_3, chlorine reacts to form chloramines, thus decreasing the level of free chlorine available for disinfection. Hence, high chlorine doses are required for the destruction and removal of ammonia, so that free chlorine will remain in water. This quantity of chlorine is called breakpoint concentration. Chlorination of water or treated effluent, known as breakpoint chlorination, is widely practiced in the water and wastewater treatment processes.[14]

Theoretically, the weight ratio of chlorine to NH_3 at the breakpoint chlorination is 7.6:1. Practically, this ratio varies between 8:1 and 10:1.

2. Reducing agents

Chlorine combines with various reducing components like H_2S, Fe^{2+}, Mn^{2+}, and NO_2^-.

$$H_2S + Cl_2 \rightarrow 2HCl + S \tag{7.30}$$

The presence of these reducing agents interferes with the use of chlorine as a disinfectant and consequently increases the demand of chlorine for disinfection.

3. Organic compounds

Certain organic compounds such as phenols react with chlorine and produce a variety of chlorinated products. These may be toxic or carcinogenic in nature like $CHCl_3$, chloroform. This is a matter of concern for human safety, therefore efforts are being made to minimize the production of chlorinated hydrocarbons.[15]

Measurement

1. Principle

Residual chlorine present as hypochlorous acid or hypochlorite ion in water or wastewater sample reacts with DPD (N, N-diethyl-p-phenylene diamine) and produces a red-colored complex. This forms the basis of the residual chlorine analytical test. The intensity of color is proportional to the concentration of residual chlorine present in the sample. The absorbance of color is measured spectrophotometrically at 530nm wavelength.

2. Apparatus

a. Spectrophotometer — Hach DR / 4000
b. Volumetric flasks — 100 and 500 ml capacity

3. Reagents

a. Hach DPD free Cl residual powder pillows
b. Hach DPD total Cl residual powder pillows
c. Stock potassium permanganate solution, $KMnO_4$ — Weigh accurately 0.4455 g $KMnO_4$ and transfer to a 500-ml volumetric flask. Dissolve in DDW. Dilute up to the mark with DDW.

$$1 \text{ ml stock } KMnO_4 \text{ solution} \equiv 1.0 \text{ mg free chlorine}$$

d. Standard KMnO4 solution: Measure 1.0 ml of Stock $KMnO_4$ solution in a 100-ml volumetric flask and dilute to the mark with DDW.

$$1\,L \text{ Standard } KMnO_4 \text{ solution} \equiv 10 \text{ mg free chlorine}$$

4. Standardization

Prepare the calibration standards of KMnO4 corresponding to 0.5, 1, 2, 3 and 4 mg/L free chlorine and standardize as follows:

a. Into five different 100-ml volumetric flasks measure 5, 10, 20, 30 and 40 ml standard $KMnO_4$ solution.
b. Dilute solution in each flask up to the mark with DDW.
c. Take 10 ml of the first standard solution in the sample cell. Add one DPD free chlorine powder pillow.
d. Swirl the sample cell for 20 seconds to mix. Wait for about 30 seconds for color development.
e. Adjust wavelength to 530 nm and place the sample cell in spectrophotometer.
f. Prepare a blank with 10-ml DDW instead of standard solution.
g. Record the absorbance of standard solution against blank at 530 nm.
h. Repeat the procedure with all standard solutions and plot a calibration curve of the absorbance against the concentration of free residual chlorine in mg/L.

5. Procedure

a. Free Chlorine Residual
   i. Take a 10-ml sample in a sample cell.
   ii. Add one DPD free Cl residual powder pillow.
   iii. Swirl to mix well for 20 seconds and wait for 30 seconds for color development.
   iv. Record the absorbance at 530 nm.
b. Total Chlorine Residual — Follow the same procedure as described above, substituting total chlorine residual powder pillows for free chlorine residual powder pillows.

6. Calculations

Compute the sample concentration directly from the reading of the instrument.

Note: *When the chlorine residual exceeds 4 mg/L, dilute the sample with DDW prior to analysis.*

7. Important Instructions

a. Analyze samples for free chlorine immediately after collection without any lapse of time. Estimation at site is preferable.
b. Do not use the same sampling bottles to collect samples for residual and total chlorine estimation.
c. Overflow the sampling bottles with the sample several times, and then cap to prevent any air entrapment.

7.5.4 Cyanide — Titrimetric Method

Source

Hydrocyanic acid and its salts represent the cyanide content in water and wastewater. This is an important and ubiquitous industrial chemical, hence it is present in many industrial effluents. It can access water systems via leakage of wastewater pipes or discharge of industrial wastes to surface waters.

Significance

Cyanide and its compounds are highly toxic, especially at low pH, i.e., under acidic conditions. Cyanide ion has the ability to coordinate with the iron (Fe) atom in a hemoglobin molecule, and thus blocks the uptake of oxygen by the blood. It acts as an inhibitor of the enzyme of phosphorylative oxidation in the respiration process. It is toxic for the aquatic ecosystem also.

Metallo-cyanide complexes formed by the reaction of CN^- with heavy metals are extremely toxic in nature. Therefore, control of cyanide in industrial effluents before discharge to a wastewater collection system is extremely important.

Ground water, surface water and the drinking-water supply, points of the seepage of industrial waste or effluent, always require closed monitoring for the presence of cyanide. Raw water used for drinking water processing must not contain more than 0.005-mg/L cyanide. Potable water must be totally free of cyanide.

Measurement

1. Principle

The cyanide (CN^-)-containing compound produces hydrocyanide (HCN) gas in the presence of H_2SO_4 during distillation. The gas is absorbed in NaOH solution and titrated with standard $AgNO_3$ to form soluble cyanide complex with silver, $[Ag(CN)_2]^-$. The end point is detected by the development of violet color with diphenyl carbazide indicator.

2. Apparatus

a. Cyanide distillation assembly — Hach model comprising a distillation flask, thistle tube, condenser, gas bubbler, absorber, flow meter, filter flask, aspirator, and a suction device. The complete unit is shown in Figure 7.1.

b. Magnetic stirrer with heater

c. Volumetric flasks — 100, 250 ml and 1 L capacity

d. Measuring cylinders — 250 ml capacity

e, Wash bottle — containing DDW

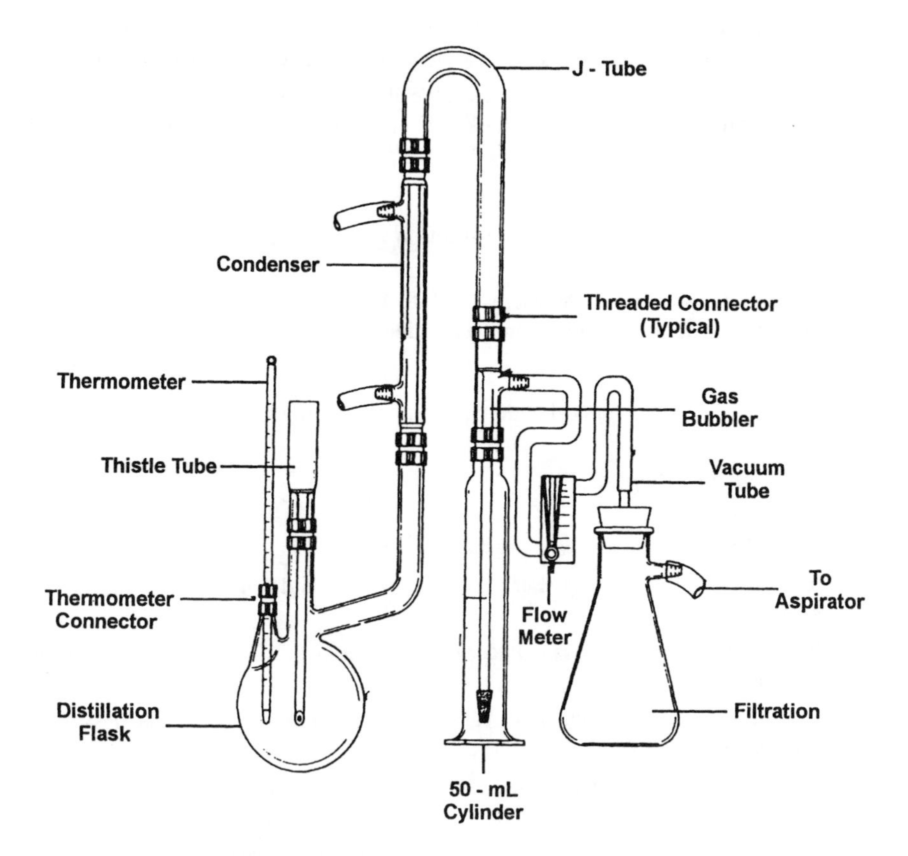

Figure 7.1 Cyanide assembly (Hach Company).

3. Reagents

a, Sodium hydroxide solutions, NaOH, 1.5 N and 6 N: Refer to Section 3.5.3 and Table 3.5.

b. Sulfuric Acid, H_2SO_4, 50% (v/v): Measure 250 ml DDW in a 1-L beaker. Carefully add 250 ml concentrated H_2SO_4 in small increments in DDW. Cool the solution to room temperature.

Note: This is an exothermic reaction, so prepare this solution with extreme care in a fume hood.

c. Magnesium chloride solution; $MgCl_2 \cdot 6\ H_2O$, 2.5 M: weigh 50.8 g $MgCl_2 \cdot 6\ H_2O$ and transfer to a 100 ml volumetric flask. Dissolve in DDW. Bring the final volume up to the mark with DDW.

d. Silver nitrate titrant; $AgNO_3$, 0.01 N: Weigh 1.699 g silver nitrate and transfer to a 1-L volumetric flask. Dissolve in DDW and dilute up to the mark with DDW.

e. Indicator solution: Weigh 0.02-g p-dimethylamino-benzalrhodanine and transfer to a 100-ml volumetric flask. Dissolve in acetone. The final volume should be 100 ml.

f. Stock potassium cyanide solution; KCN: Weigh 0.251 g KCN and 0.5 g KOH. Transfer the mixture to a 1-L volumetric flask. Dissolve it in DDW. Add 5 ml of 6N NaOH solution. Dilute up to the mark with DDW.

$$1 \text{ L stock cyanide solution} \equiv 100 \text{ mg } CN^-$$

g. Standard cyanide solution: Measure 10-ml stock cyanide solution in 100-ml volumetric flask. Add 1 ml of 6N NaOH solution and dilute with DDW up to the mark.

$$1 \text{ L standard cyanide solution} \equiv 10 \text{ mg } CN^-$$

4. Distillation of Sample

a. Set up the distillation apparatus as shown in Figure 7.1 in a fume hood.

b. Turn on the circulating tap water and make sure it is flowing steadily through the condenser.

c. Fill the distillation apparatus cylinder to the 50-ml mark with 1.5 N NaOH standard solution.

d. Measure 250-ml sample with a clean graduated cylinder into the distillation flask. Place a stirring bar in the flask. Attach the thistle tube.

e. Do not connect the vacuum tubing to the gas bubbler at this stage. First turn on the water to the aspirator for smooth flow and adjust the flow meter to 0.5 SCFH.

f. Connect the vacuum tubing to the gas bubbler to maintain a continuous airflow (check the flow meter). Make sure that air is bubbling from the thistle tube and the gas bubbler.

g. Turn the power switch on and set the stir control to position 5.

h. At this stage, add 50 ml of 50% (v/v) H_2SO_4 solution through the thistle tube into the distillation flask with care.

i. Rinse the thistle tube with a small amount of DDW using a wash bottle.

j. Stir the mixture for 3 min thoroughly, then add 20-ml $MgCl_2$ solution through thistle tube. Rinse again with DDW. Stir the mixture for 3 min more.

k. Before starting the heating control, maintain the constant flow of water through the condenser. This is an essential step.

l. Turn the heat control to position 10.

m. Monitor the distillation flask carefully at this stage. Once the sample starts boiling, reduce the airflow and set to position 0.3 SCFH.

Note: If the contents of distillation flask begin to rise up in the thistle tube increase the air flow by adjusting the flow meter until the contents flow back to the distillation flask through the thistle tube.

n. Reflux the contents for 1 hr.

o. When 1 hr refluxing is over, turn off the heating mantle but maintain the airflow for 15 min more.

p. After 15 min, remove the rubber stopper on the vacuum flask to break the vacuum and turn off the water to the aspirator.

q. Turn off the water to the condenser.

r. Remove the gas bubbler / absorber assembly from the distillation apparatus.

s. Pour the contents of the cylinder into a 250-ml volumetric flask. Rinse the gas bubbler, cylinder and J tube with DDW and add washings to the volumetric flask.

t. Fill the flask up to the mark with DDW and mix thoroughly. Analyze this solution for cyanide content.

5. Standardization

Prepare calibration standards containing 0.05, 0.10, 0.50 and 1.0 mg CN/L and standardize as follows:

a. Into four different 100-ml volumetric flasks measure 0.5, 1.0, 5.0 and 10 ml standard cyanide solution.

b. Dilute solution in each flask up to the mark with DDW and invert several times to mix thoroughly.

c. Add 0.5 ml of indicator solution to each flask.

d. Titrate with standard $AgNO_3$ titrant to the first change in color from canary yellow to a brownish pink.

e. A blank is prepared with 100 ml DDW instead of standard solution containing 1ml of 6 N NaOH solution. Titrate the blank in the same manner as described for the standards.

6. Procedure

a. Dilute the distilled solution containing cyanide so that it will require less than 10 ml of $AgNO_3$ titrant for titration. Take 100 ml of diluted sample.

b. Add 0.5 ml of indicator solution.

c. Titrate with standard $AgNO_3$ titrant. Change in color from canary yellow to a brownish pink indicates the end point.

7. Calculations

$$\text{mg CN}^-/\text{L} = \frac{(\text{A} - \text{B}) \times 1000}{\text{volume of sample, ml}} \times \begin{matrix}\text{dilution factor} \\ \text{(if required)}\end{matrix}$$

where A = volume of titrant used for the sample, ml
B = volume of titrant used for the blank, ml

8. Important Instructions

a. The pH of all stock and standard solutions should fall in alkaline range. Therefore, add 6 N NaOH at the rate of 5–8 ml/L of cyanide solution to keep it alkaline.
b. Set up the assembly in a fume hood because cyanide and hydrogen cyanide gas (HCN) are both extremely poisonous.
c. Before starting the heating system in the distillation unit, a steady flow of water through the condenser should be achieved.

7.5.5 Fluoride — SPANDS Method — Detectable Range: 0 to 2.0 mg/L fluoride

Fluoride, the most commonly occurring form of fluorine, is the natural mineral contaminant of water. Its quantity in water must be controlled to maintain human health. Therefore, personnel engaged in water management and treatment activities have dual responsibilities:

a. To design and operate the water treatment process units for the removal of excessive fluorides if the natural water has higher concentration than permitted.
b. To supervise and monitor the addition of fluorides to achieve the optimum level in water supplies in the areas having concentrations lower than the recommended levels for human consumption.

Source

1. Rocks

Fluoride is generally found in nature in a few types of igneous and sedimentary rocks.

2. Ground water

Usually contains fluoride dissolved by geological formations.

3. Surface waters and industrial wastes

Contain appreciable amounts of fluoride, especially in the areas near aluminum-processing plants. These plants use cryolite (Na_3AlF_6) as a solvent in the Al-extraction process. Therefore, an appreciable amount of fluoride escapes into the environment through exhaust vents. Also, the wastes of these plants contain a significant concentration of fluoride.

Significance

1. Dental disease — fluorosis or mottling

Fluoride in large quantities is harmful for humans and other animals, but is beneficial if present in a low concentration. Excessive consumption of fluorine through drinking water causes a dental disease known as fluorosis or mottling. This causes discoloration of teeth when fluoride concentrations exceed 2.0 mg/L in potable water.[16] Excessive regular consumption of fluorine may give rise to bone fluorosis, i.e., deformation of bones and other skeletal disorders.

2. *Beneficiary aspects*

A concentration of approximately 1.0 mg/L in drinking water helps to prevent dental cavities in children.[17] During tooth formation, fluorine combines with tooth enamel, which results in the formation of hard, strong teeth that have a greater resistance to decay.

The maintenance of an optimum level of fluorine in water supplies is significant for public health. Thus, estimation of fluorine in water is an extremely important factor. In areas where fluorine is added to maintain an optimum level for the control of dental diseases, it is essential to first determine the amount of natural fluorine present in water. This will help assess the appropriate amount of fluorine to be added to the water supply to achieve the optimum level.

Measurement

1. *Principle*

The colorimetric method of fluoride determination is based on the reaction of fluoride with the zirconium ion present in red-colored SPANDS dye, which results in the formation of a colorless complex (ZrF_6^{2-}). The extent of bleaching of the red color is proportional to the amount of fluoride present in the water sample.

2. *Interference*

The presence of high levels of alkalinity at 5000 mg/L, chloride 7000 mg/L, free chlorine <2 mg/L, sulfate 200 mg/L and iron 10 mg/L may cause interference in the estimation of fluoride ion in water samples, especially in wastewater. Hence, prior to laboratory analysis, the sample is distilled to remove this interference.

3. *Apparatus*

a. Distillation unit — The complete distillation assembly comprising a distillation flask, J-tube condenser, collection flask and drip, graduated cylinder, and 100-ml tube is shown in Figure 7.2
b. Thermometer
c. Magnetic stirrer with heater
d. UV light Spectrophotometer — any commercially available model set at 580-nm wavelength.
e. Beakers — 50 and 500 ml capacity
f. Volumetric flasks — 500 ml and 1-L capacity
g. Cylinders — 100 and 25 ml capacity

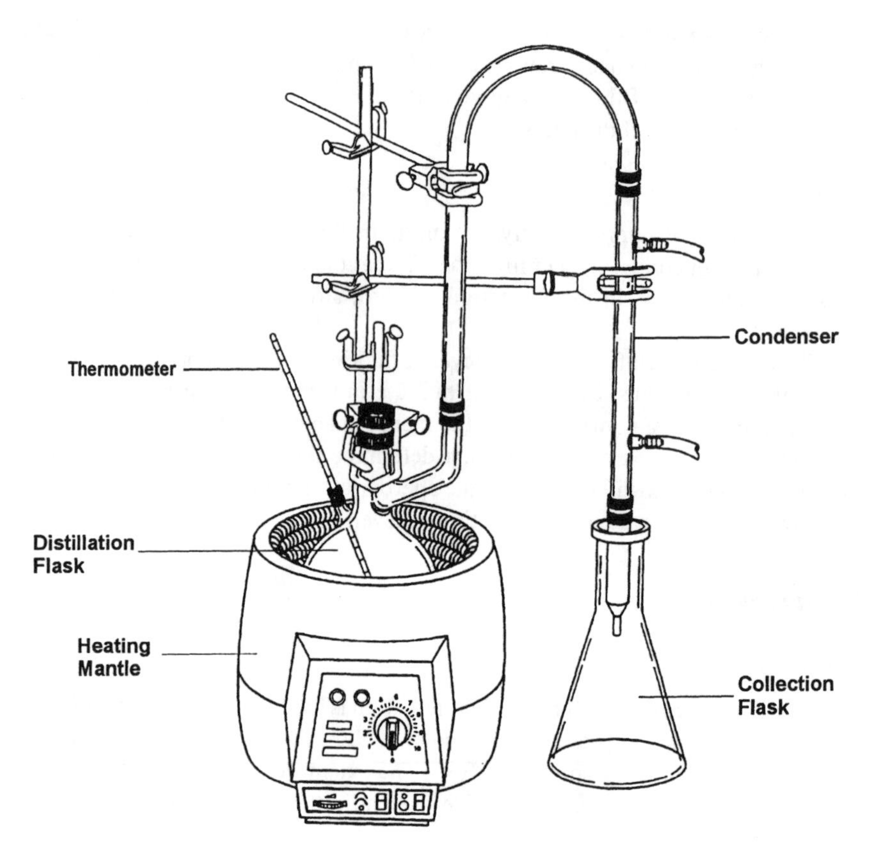

Figure 7.2 Distillation unit (Hach Company).

4. Reagents

a. Sulfuric acid, H_2SO_4, 2:1 solution: Measure 100 ml DDW in a 500-ml beaker. Keep it on a magnetic stirrer. Slowly and carefully add 200 ml concentrated H_2SO_4 into DDW with constant stirring. Cool the solution to room temperature.

Note: Prepare this solution in a fume hood.

b. SPANDS solution: sodium 2-(p-sulfophenyl azo)-1,8-dihydroxy-3, 6- naphthalene disulfonate, also known as 4,5-dihydroxy-3-(p-sulfophenyl azo)-2,7-naphthalene disulfonic acid, tri-sodium salt: weigh 1.916 g SPANDS and transfer to a 1-L volumetric flask. Dissolve in DDW. Dilute up to the mark with DDW.

Note: This solution is highly stable and can be stored for a long period.

c. Zirconyl-acid reagent: Zirconyl chloride octahydrate, $ZrOCl_2.8H_2O$: Weigh 0.133 g of this compound and transfer to a 500 ml volumetric flask. Dissolve in a small volume of DDW. Add 350 ml concentrated HCl and dilute up to the mark with DDW.

d. SPANDS - zirconyl acid reagent: Measure 500 ml of SPANDS in a 1-L volumetric flask. Add 500 ml of Zirconyl-acid reagent to the flask. Mix the contents thoroughly to prepare SPANDS–Zirconyl-acid reagent. This reagent remains stable for a period of 2 years.

e. Dechlorinating agent, sodium arsenite, $NaAsO_2$ solution: Weigh 0.5-g sodium arsenite and transfer to a 100-ml volumetric flask. Dissolve in DDW. Dilute up to the mark with DDW.

f. Reagent blank: Measure 10 ml SPANDS reagent in a 100-ml volumetric flask. Add 50 ml DDW to the flask. Now add 7 ml concentrated HCl in the flask. Make up the volume to 100 ml with DDW. Use this solution to set the spectrophotometer at Zero.

g. Sodium fluoride stock solution, NaF: Weigh 0.1105-g anhydrous NaF and transfer to a 500-ml volumetric flask. Dissolve in DDW. Dilute to 500 ml with DDW.

$$1\text{ L stock fluoride solution} \equiv 100\text{ mg }F^-$$

h. Standard fluoride solution: Measure 10.0-ml stock fluoride solution in a 100-ml volumetric flask. Dilute up to the mark with DDW.

$$1\text{ L standard fluoride solution} \equiv 10\text{ mg }F^-$$

5. Preparation of the sample — distillation process

a. Set up the distillation apparatus as shown in Figure 7.2.

b. Turn on the water and maintain continuous flow through the condenser.

c. Measure 100 ml of the sample into the distillation flask using a 100-ml graduated cylinder.

d. Turn the stirrer power switch on and adjust the control to position 5.

e. Add carefully 250 ml 2:1 mixture of concentrated H_2SO_4 and water to the distillation flask with utmost care.

f. Place the thermometer through the opening and turn the heater on and set at position 10. At this stage, yellow indicator lamp illuminates.

g. When the temperature reaches 180°C or when 100 ml of the distillate has been collected, switch off the heater and stirrer both. The distillation of sample will be completed in about 1 h.

h. Dilute the distillate, if required, and use it for estimation of fluoride content.

6. Standardization

Prepare calibration standards containing 0.1, 0.5, 1.0, and 1.5 mg/L F^- from standard fluoride solution and standardize as follows:

a. Into four different 100-ml volumetric flasks, measure 1, 5, 10, and 15 ml of 10-mg/L standard fluoride solution.

b. Dilute solution in each flask up to the mark with DDW and mix thoroughly.

c. Into four different 50-ml beakers, measure 25-ml of above-mentioned calibration solutions and add 5.0 ml of SPANDS–Zirconyl-acid reagent to each.
d. Set the wavelength at 580 nm.
e. Set the spectrophotometer at zero using the reagent blank instead of DDW.
f. Record the readings and prepare a calibration curve of absorbance against the concentration of F^- mg/L.

7. Procedure

a. If the sample contains residual chlorine, add 0.1 ml of dechlorinating agent for each 0.1-mg free chlorine and mix the solution thoroughly.
b. Place 25 ml sample or an aliquot diluted to 25 ml in spectrophotometric cell and add 5.0 ml acid - Zirconyl - SPANDS reagent.
c. Read the absorbance at 580 nm.

8. Calculation

Record the concentration of F^- in mg/L in a sample directly from the linear calibration curve prepared during standardization of the procedure.

If the sample was diluted, multiply the concentration obtained from standard curve by dilution factor.

9. Important Instructions

a. The basic of fluoride estimation is the decoloration of the reagent, so the addition of SPANDS must be done with high accuracy, as a small variation in reagent quantity could cause serious errors in the estimation.
b. SPANDS reagent contains sodium arsenite, so the final solutions will contain a considerable amount of arsenic. These solutions should be disposed of with care, as described in Section 2.2.1.b.
c. The standard curve will be an inverse curve. The blank will have higher absorbance than the samples at 580 nm.

7.5.6 Silicate — Silicomolybdate Method

Source

Silica is found in abundance in the earth's crust, therefore all natural waters contain silica in significant quantities. Normally it ranges between 1–30 mg/L but sometimes may be as high as 100-mg/L.[18]

Significance

The presence of silica in natural waters is undesirable because it restricts the use of water for industrial purposes. Silica combines even with trace amounts of Ca, Mg,

Al and other metallic constituents of water to form objectionable silicate scales in heating equipment, specially boilers. These scales are difficult to remove. Hence, first removing the silica from water is essential. At high temperature, silica becomes highly volatile, and, on cooling, forms a hard glass-like coating on turbine blades.

Measurement

1. Principle

Silica available in the sample reacts with molybdate ion under acidic conditions, i.e., at pH between 1 and 2 and produces yellow silicomolybdic acid. Silica is then determined by measuring the intensity of yellow color of silicomolybdic acid spectrophotometrically at 452 nm.

2. Interference

Phosphates present in the sample interfere with this reaction, which produces phosphomolybdic acid complexes with molybdate ion. Therefore, oxalic acid is used to remove this interference.

3. Apparatus

a. Visible and UV Spectrophotometer — set at 452 nm
b. Heater with magnetic stirring
c. Volumetric flasks — 100-ml capacity
d. Beaker — 250-ml capacity

4. Reagents

a. Hydrochloric acid, HCl 50% (v/v): Measure 500 ml DDW in a 1-L volumetric flask. Slowly add 500 ml concentrated HCl to DDW. Cool the solution to room temperature.
b. Sodium hydroxide, NaOH, 1 N: Refer to Section 3.5.3 and Table 3.5.
c. Ammonium molybdate reagent; $(NH_4)_6Mo_7O_{24}.4H_2O$:
   i. Weigh 10.0 g ammonium molybdate and transfer to a 250-ml beaker. Dissolve in DDW.
   ii. Keep the beaker on a heater equipped with magnetic stirring. Turn on the heater and stirrer. Dissolve the salt in DDW with constant stirring and warming gently.
   iii. After complete dissolution, transfer the solution to 100-ml volumetric flask. Make up the volume to 100 ml with DDW.
   iv. Filter any remaining residue.
   v. Adjust pH to 7–8 with 1N NaOH for stabilization.
d. Oxalic acid, $H_2C_2O_4.2\ H_2O$, 2 N: Weigh 12.6 g oxalic acid and transfer to a 100-ml volumetric flask. Dissolve in DDW. Dilute up to the mark with DDW.
e. Reagent blank: Measure 100 ml DDW in a 100-ml volumetric flask and transfer to a 150-ml Erlenmeyer flask. Add 1.0 ml of 50% (v/v) HCl solution and 2.0 ml ammonium

molybdate solution. Mix thoroughly by inverting the flask several times. Now add 1.5-ml oxalic acid and mix again. Use this reagent blank to adjust spectrophotometer at zero.

f. Silica stock solution: Purchase this solution containing 1000-mg silica/L readymade from Hach Co.

g. Silica standard solution: Measure 10 ml of silica stock solution in a 100-ml volumetric flask and dilute to the mark with DDW.

$$1 \text{ L silica standard solution} \equiv 100 \text{ mg silica}$$

Note: *Store all reagents in polyethylene containers. Avoid using glass apparatus as much as possible during the estimation procedure.*

5. Standardization

Prepare calibration standards containing 5, 10, 20, 30 and 50 mg/L silica and standardize as follows:

a. Into 5 different 100-ml volumetric flasks, pipette 5, 10, 20, 30 and 50 ml of silica standard solution.
b. Dilute solution in each flask to marked volume with DDW.
c. Add 1.0 ml of 50% (v/v) HCl and 2.0 ml of ammonium molybdate solution to each flask.
d. Mix well by inverting the flasks several times. Allow the solution to stand for 5–10 min.
e. Add 1.5 ml oxalic acid and mix thoroughly. Allow the solution to stand for two more minutes.
f. Use reagent blank and set spectrophotometer at zero.
g. Record the absorbance of color developed in the standard solutions against the reagent blank at 452 nm.
h. Plot a calibration curve of absorbance against concentration of silica in mg/L.

6. Procedure

a. Take 100-ml sample or an aliquot quantity diluted to 100 ml and follow the same procedure of color development as described under standardization.
b. Record the absorbance of the sample against reagent blank at 452 nm.

Note: *Always prepare fresh reagent blank with new lot of reagents and read its absorbance against DDW.*

c. Record the absorbance against the reagent blank.

7. Calculation

Read the concentration of silica in mg/L directly from the calibration curve. If the sample is diluted, the reading must be multiplied by the dilution factor.

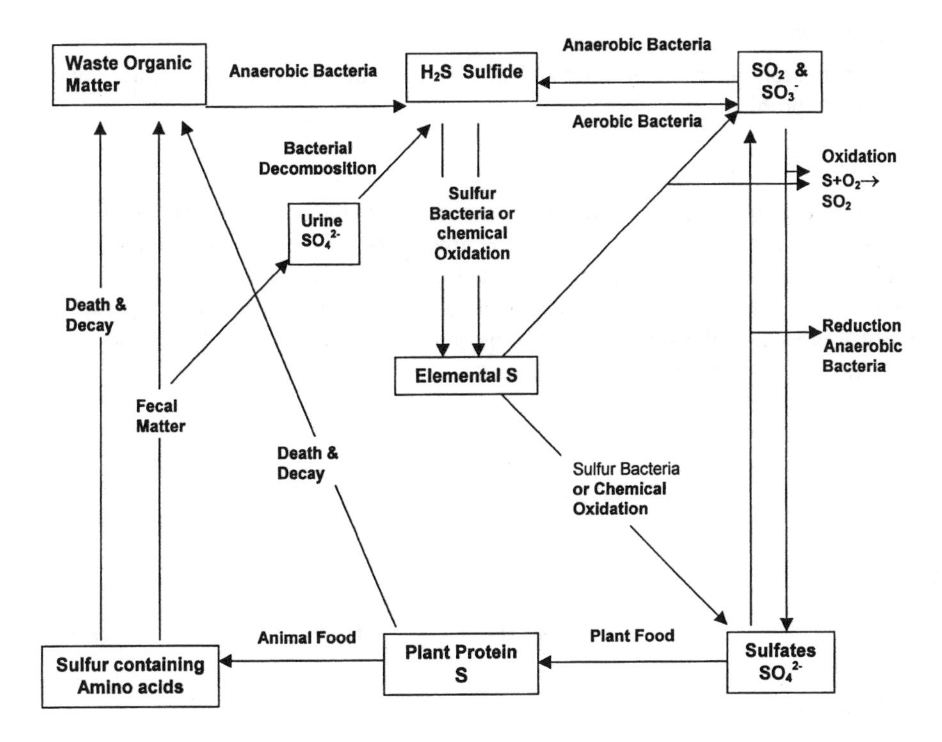

Figure 7.3 Sufur cycle

7.5.7 Sulfate — Turbidimetric Method — Minimum Detectable Limit: 1 mg/L SO_4^{2-}

Sulfur is an essential element of all the three parts of the biosphere, i.e., land (lithosphere), water (hydrosphere) and air (atmosphere) because it exists in solid, gaseous as well as in soluble form. The transformation of sulfur into different organic and inorganic products is shown in the sulfur cycle, Fig. 7.3.

Sulfate is the most common product of the sulfur cycle in nature.

Source

1. Natural waters

Since the sulfate (SO_4^{2-}) ion is one of the most universal anions that occur in natural waters, it appears in a significant concentration in surface, ground and marine waters.

2. Domestic wastewater

Sulfur is present in human excreta in the form of sulfur-containing amino acids. Therefore, domestic wastewater is rich in sulfur. It is available in the form of organic and inorganic sulfur compounds such as hydrogen sulfide (H_2S), mercaptans, thioesters and disulfides.

Significance

The estimation of sulfate in water and wastewater is of great concern because of the following problems:[19]

1. Odor Generation

Wastewater collection and treatment is connected with the production of malodorous volatile compounds like H_2S, mercaptans, thioesters etc. The major problem exists with the generation of H_2S. Sulfate-reducing bacteria decompose organic matter under anaerobic conditions producing malodorous compounds at the expense of oxygen present in the sulfate anion. Thus, sulfates are reduced to H_2S as shown in Equations 7.31 and 7.32:

$$SO_4^{2-} + \text{organic matter} \xrightarrow[\text{bacteria}]{\text{sulfate reducing}} S^{2-} + H_2O + CO_2 \tag{7.31}$$

$$2\ H^+\ (\text{from } H_2O) + S^{2-} \rightarrow H_2S \tag{7.32}$$

2. Sewer Corrosion

This is another serious problem associated with the presence of sulfate in wastewater. The H_2S produced in wastewater is oxidized to H_2SO_4 in the atmosphere below the crown of manholes and sewer pipes. This action is carried out by sulfide oxidizing bacteria mainly of the thiobacillus genus, as shown in Equation 7.33:

$$H_2S + 2\ O_2 \xrightarrow{\text{thiobacillus}} H_2SO_4 \tag{7.33}$$

Acid thus produced causes corrosion in sanitary sewers.

3. Health Hazards:

a. Hydrogen sulfide is a highly toxic and health-hazardous gas. There is evidence that a long exposure to a concentration of 0.03% (300 ppm) of H_2S in the air has caused death.[9] Concentrations exceeding 1000 ppm can be fatal if humans are exposed for even a few min.[20]

b. Sulfate levels of 2000 mg/L have been found to cause progressive weakening and death in cattle.[21]

4. Scale formation

The level of sulfate should be monitored in public and industrial water supplies because of its tendency to form hard scales in boilers and heat-dissipation systems.

Measurement

1. Principle

SO_4^{2-} ions in water or wastewater sample have a tendency to precipitate out as $BaSO_4$ on reaction with barium chloride ($BaCl_2$) under acidic conditions. This tendency of sulfate ions increases in the presence of a conditioning agent. The amount of precipitation is proportional to the concentration of SO_4^{2-} ions in the sample.

2. Apparatus

a. U V spectrophotometer — set at 420 nm
b. Magnetic stirrer
c. Oven — set at 103°C
d. Desiccating cabinet
e. Filtration unit — as described in Section 6.6.1.2a
f. Glass fiber filter papers — GF/C or equivalent
g. Volumetric flasks — 100, 500 ml and 1 L capacity
h. Beakers — 250 ml capacity

3. Reagents

a. Conditioning reagent:

i. Hydrochloric acid, HCl — Concentrated, add 30 ml concentrated HCl to 300 ml DDW in a 500-ml volumetric flask.
ii. Ethyl alcohol, 95% - Add 100 ml, 95% ethyl alcohol in the flask.
iii. Sodium chloride, NaCl — Weigh 75 g NaCl and add to the above solution.
iv. Glycerol — Measure 50 ml glycerol and add to the above solution.

Mix thoroughly the contents of 500-ml volumetric flask. Make up the final volume to 500 ml with DDW. This solution is known as the conditioning reagent.

b. Barium chloride solution $BaCl_2 . 2\ H_2O$: Weigh 100 g $BaCl_2 . 2\ H_2O$ and transfer to a 1-L volumetric flask. Dissolve in DDW. Dilute to 1 L with DDW. Filter through a glass fiber filter GF/C.

$$1.0 \text{ ml of } BaCl_2 \text{ solution precipitates} \approx 40 \text{ mg } SO_4^{2-}$$

c. Standard sulfate solution, sodium sulfate, Na_2SO_4:

i. Dry an adequate quantity of anhydrous Na_2SO_4 in an oven set at 103°C for 1 h. Cool it to room temperature by keeping in a desiccating cabinet.
ii. Weigh 0.1479 g dry and cool anhydrous Na_2SO_4 and transfer to a 1L volumetric flask and dissolve in DDW. Dilute to 1 L with DDW.

$$1 \text{ L standard sulfate solution} \equiv 100 \text{ mg } SO_4^{2-}$$

4. *Sample Preparation*

a. Colored or suspended material in water or wastewater sample will interfere and produce false readings. These should be removed prior to sulfate measurement as described below:

   i. The suspended material should be removed by filtration. Pass the sample through a filtration unit using GF/C filter paper.

   ii. If both colored and suspended material are in minute quantities and cannot be removed by filtration, prepare a sample blank without $BaCl_2$. Set the spectrophotometer at zero with this blank before proceeding for sulfate measurement.

b. If sample has sulfate concentration more than 50 mg / L, dilute it.

5. *Standardization*

For calibration prepare standard solutions containing 5, 10, 20, 30, 40 and 50 mg/L sulfate and standardize as follows:

a. Into six different 100-ml volumetric flasks measure 5, 10, 20, 30, 40 and 50 ml of sulfate standard solution.

b. Dilute solution in each flask up to the mark with DDW. Transfer these solutions in six different beakers of 250-ml capacity.

c. Keep the beakers on magnetic stirrer. Add 5 ml conditioning agent to each beaker with constant stirring.

d. Add 5 ml $BaCl_2$ solution slowly with gentle stirring.

e. Prepare a reagent blank using DDW instead of standard solution. Add 5 ml conditioning reagent and 5 ml $BaCl_2$ solution in it.

f. Adjust the wavelength of spectrophotometer at 420 nm. Set zero with reagent blank.

g. Read the absorbance against reagent blank. The reading should be taken within 4 min.

h. In the presence of colored or suspended material read the absorbance of sample against a sample blank.

i. Plot a calibration curve of absorbance vs. concentration of sulfate in mg/L.

6. *Procedure*

a. Take 100-ml sample or an aliquot diluted to 100 ml into a 250-ml beaker.

b. Add 5 ml conditioning agent and 5 ml $BaCl_2$ solution to the sample. If the addition of last 2-ml $BaCl_2$ solution does not produce any noticeable precipitate, the precipitation is considered complete. If the sample consumes more $BaCl_2$ solution, dilute it.

c. Follow the same procedure as described under Standardization.

d. Read the absorbance.

7. Calculation

Calculate sulfate concentration by comparing turbidity reading with a calibration curve or by direct readout of the instrument. If the sample is diluted, the result must be multiplied by dilution factor.

7.5.8 Sulfide — Dissolved and Total — Iodometric Method

The generation and emission of reduced sulfur components, mainly volatile hydrogen sulfide gas, is the major cause of odors in sanitary sewers. The emission of gaseous hydrogen sulfide in the environment causes a nuisance to the public because of its rotten-egg smell and health-hazardous nature.

For source and significance, refer to Section 7.5.7.

Microbial Production of Sulfide in Sewerage Systems

1. Basic Definitions

The following terms are essential to understanding the basic concepts of biological and chemical processes that lead to the generation and build up of hydrogen sulfide in sewers:

a. Hydrogen Sulfide (H_2S)

Hydrogen sulfide is a gaseous component that occurs both in the sewer atmosphere and as a dissolved species in wastewater. This gas imparts a rotten-egg odor to the wastewater. It is highly toxic and hazardous to human health. It is oxidized by the action of bacteria on exposed sewer surfaces to form sulfuric acid, the major cause of corrosion. On decomposition, it produces hydrogen sulfide (HS^-) and H^+ ions.

$$H_2S \leftrightarrow H^+ + HS^- \tag{7.34}$$

b. Hydrogen Sulfide Ion (HS^-)

This can be formed by the reversible dissociation of dissolved hydrogen sulfide as shown in Equation 7.34. It is the combination of a positive hydrogen ion and a bivalent negative sulfide ion.

c. Sulfide Ion (S^{2-})

This is formed by second dissociation of aqueous hydrogen sulfide.

$$HS^- \leftrightarrow H^+ + S^{2-} \tag{7.35}$$

d. Sulfide

A general term that can refer to any or all species containing the sulfide ion as mentioned above.

2. Basic Reactions

Wastewater consists of both organic and inorganic matter and different species of aerobic and anaerobic microbes that decompose organic matter in the presence and absence of oxygen respectively through the following three main steps:

i. Catabolism — the breakdown of carbon substrate in an anaerobic environment during which a small amount of ATP (adenosine triphosphate) is formed. This is an energy-rich molecule. Primary electron donors like NAD (nicotinamide adenine dinucleotide), NADP (nicotinamide adenine dinucleotide phosphate) or flavoproteins are reduced in this process.

ii. Electron Transport — an anaerobic process that leads to the generation of ATP (respiratory chain phosphorylation).

iii. Oxidation — a reaction where free or combined oxygen is reduced to OH^- by cytochrome oxidase.

Catabolism and electron transport are similar processes in both aerobic and anaerobic decomposition of organic matter. The major difference lies in the last step from aerobes to anaerobes. As free oxygen is consumed for oxidation in aerobic decomposition, under anaerobic conditions combined oxygen available in the form of sulfate anion is utilized for this purpose. Sulfate anion itself is reduced to Sulfide. The end product of the aerobic process is an oxidized product, i.e., SO_4^{2-} ion. That of an anaerobic process is a reduced product, i.e., S^{2-} ion. The sequences of aerobic and anaerobic microbial decomposition of organic matter are presented in Figures 7.4 and 7.5.

3. Processes of Generation of Hydrogen Sulfide (H_2S) in Sewers

The generation of H_2S is the major problem associated with wastewater collection and treatment facilities. The following two processes are involved in the generation of H_2S:

i. Sulfate Reduction: The sulfate anions is one of the universal anion occurring in natural ground water entering the sewers through infiltration. This SO_4 anion serves as a hydrogen acceptor in the absence of free O_2 and nitrate (NO_3) for biochemical oxidation of organic matter by obligate anaerobic bacteria. The most common species is *desulfovibrio*, which degrades organic matter according to Equations 7.36 and 7.37.

$$SO_4^{2-} + \text{organic matter} \xrightarrow{\text{Desulfovibrio}} S^{2-} + H_2O + CO_2 \tag{7.36}$$

$$2\,H^+ + S^{2-} \rightarrow H_2S \tag{7.37}$$

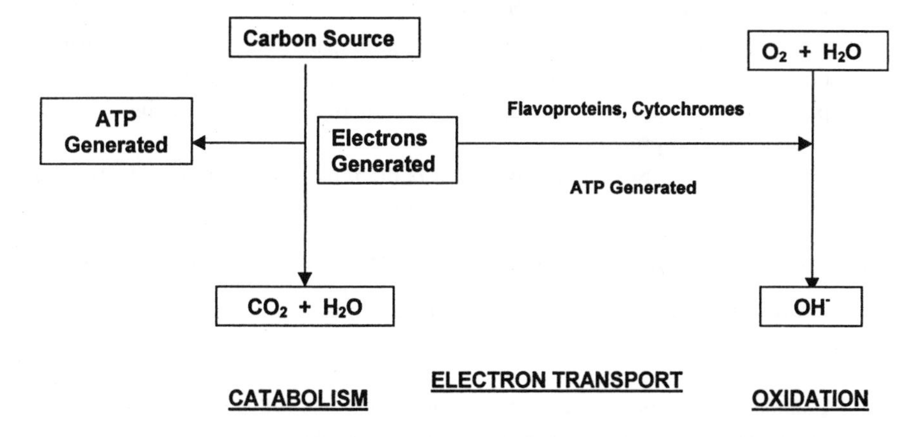

Figure 7.4 - The schematic representation of aerobic microbial decomposition of organic matter (adapted from Postgate, J. R., *The Sulfate Reducint Bacteria*, 2nd ed., Cambridge University Press, 1984, with permission)

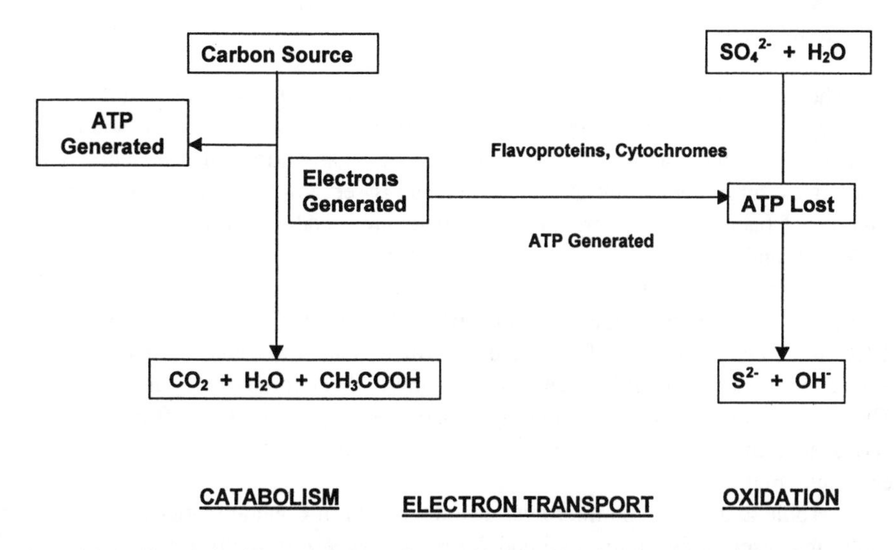

Figure 7.5- The schematic representation of anaerobic microbial decomposition of organic matter (adapted from Postgate, J. R., *The Sulfate Reducint Bacteria*, 2nd ed., Cambridge University Press, 1984, with permission)

These microbes are commonly found both in the digestive tracts of humans and animals and in mud containing organic matter. Therefore, they are available in abundance in domestic wastewater.

ii. Reduction of Sulfur-Containing Organic Compounds

Anaerobic reduction of sulfur containing amino acids such as cysteine, cystine and methionine is another source of H_2S generation in domestic sewage. This decomposition is carried by many species of proteolytic bacteria including veillonella, clostridia and proteus.

Out of these processes sulfate reduction is the most significant mechanism for the generation of malodorous hydrogen sulfide in wastewater.

4. Common species of Microbes Producing H_2S in Waste water

The microbes that reduce sulfate to sulfide in wastewater fall under the following three categories:

i. Assimilatory bacteria: These microbes utilize inorganic sulfur for their growth and activity and reduce it to sulfide within their protoplasm.
ii. Proteolytic bacteria: Such microorganisms have the capability of hydrolyzing sulfur-containing proteins and amino acids under anaerobic conditions and releasing sulfides in the environment.
iii. Sulfate-reducing bacteria: These are strict anaerobic bacteria, which utilize inorganic sulfate as the oxygen source to oxidize organic matter. The sulfate anion itself reduces to sulfide.

The common species of sulfate reducing bacteria in wastewater is desulfovibrio desulfuricans. These are obligate anaerobes that utilize sulfate as the source of oxygen or hydrogen acceptor. Different kinds of organic molecules including carbohydrates, amino acids, lipids and organic acids serve as a carbon source for the survival of bacteria. These organic compounds are metabolized according to reactions 7.36 and 7.37.

Most of the microbial population occurs in the biofilm, i.e., the slime layer formed on the walls of a pipeline or in deposits of sludge and silt on the pipe invert. This slime layer consists of a mixed growth of anaerobic and aerobic bacteria embedded in the matrix of filamentous organisms and gelatinous material (zoogleae). The typical thickness of the slime layer is in the range from 0.3–1.0-mm[19] but it varies according to the velocity of water, suspended matter in wastewater, and environmental conditions.

Aerobic and anaerobic microbial decomposition of organic matter in the slime layer are presented in Figures 7.6 and 7.7 respectively. Aerobic processes consume free oxygen available in wastewater in the form of dissolved oxygen (DO) for oxidation (Figure 7.6). Therefore, available DO starts depleting as sewage flows down the sewer. Some oxygen is dissolved in the flowing wastewater through exposed water surface but it is not sufficient to meet the demand of aerobic decomposition. Thus, there is a stage when wastewater is completely devoid of oxygen. This condition accelerates the action of anaerobic bacteria, which reduces sulfate to sulfide. Anaerobic bacteria satisfy its oxygen demand from sulfate anion, as explained in Figure 7.7. Thus, an anaerobic environment favors the generation of hydrogen sulfide in sewers.

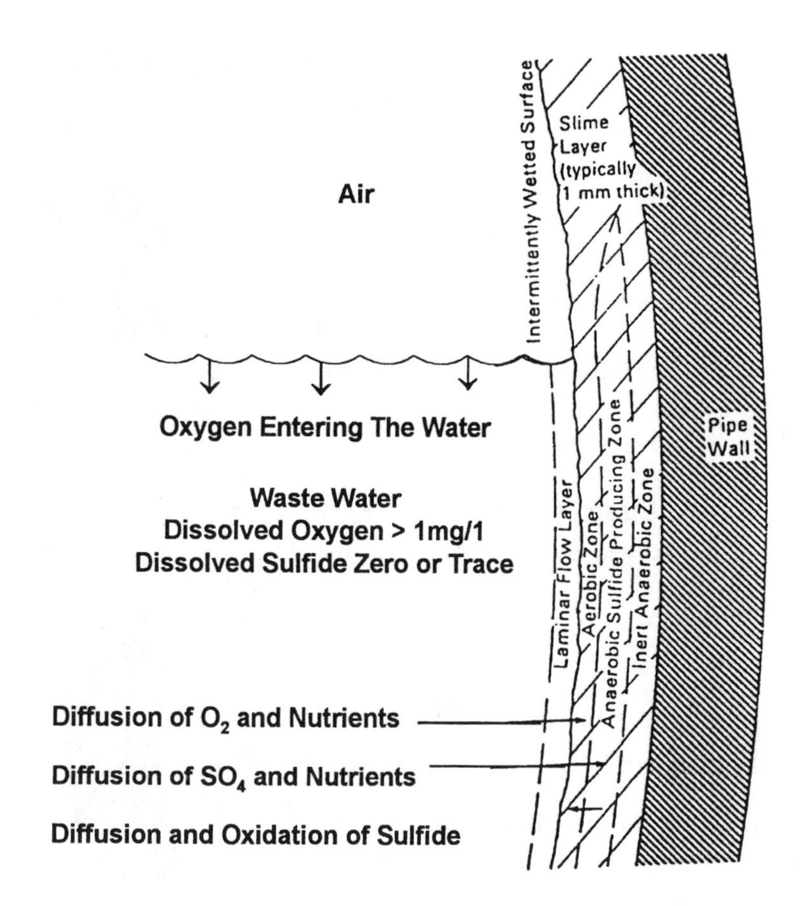

Figure 7.6 - Microbial decompostion of organic matter in slime layer under aerobic environment (adapted from EPA Design Manual: odor and corrosion control in sanitary sewerage systems and treatment plants, EPA / 625 / 1-85 / 018, USA, 1985

When sufficient oxygen is available in dissolved form in wastewater, the sulfate-reducing bacteria remain in a dormant state. There is no production of H_2S (Figure 7.6). However as soon as anaerobic conditions develop in sewers, these bacteria start reduction of sulfate ions using oxygen for oxidation. This results in the production of malodorous hydrogen sulfide.

Processes to control generation and emission of H_2S in sewerage system

Descriptions of various odor-control processes such as injection of hydrogen peroxide, NaOH, NaOCl, iron salts, sodium nitrate etc., as well as activated carbon filters, wet scrubbing system, pH control, and air / oxygen injection are available in literature.[21-25]

In 1994 this author reported a new method of odor control that included a combination of chemicals to control the generation and emission of malodorous hydrogen sulfide in wastewater.[26]

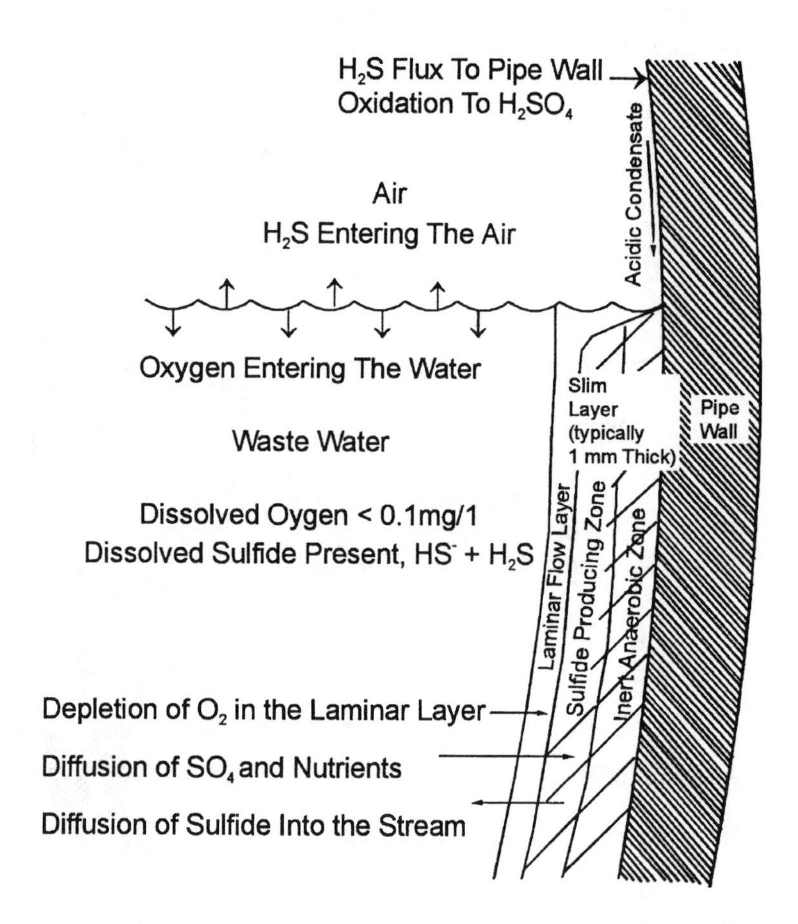

Figure 7.7 - Microbial decompostion of organic matter in slime layer under anaerobic environment (adapted from EPA Design Manual: odor and corrosion control in sanitary sewerage systems and treatment plants, EPA / 625 / 1-85 / 018, USA, 1985

Measurement

1. Sampling

Dissolved hydrogen sulfide emits through the surface of water to the atmosphere freely, so the following precautions must be taken during the collection of samples:

a. Collect samples in air-tight, narrow-mouth polyethylene bottles.

b. Collect samples without the entrapment of air bubbles with great care and keep at 4°C during transportation to the laboratory for analysis.

c. Preserve the sample immediately with zinc acetate and keep at 4°C, if total sulfide is required to estimate. Without preservation, the sample must be analyzed within half an hour of collection.

2. Principle

The suspended matter in the sample is removed by coagulation with aluminum chloride ($AlCl_3$), which, in the presence of NaOH, produces a floc of aluminum hydroxide. The floc settles, leaving behind a clear supernatant. Precipitate out dissolved sulfide as zinc sulfide (ZnS) with zinc acetate in alkaline medium. Dissolve the precipitate of ZnS in HCl and measure sulfide iodometrically.

3. Apparatus

a. Sampling bottles — Polyethylene narrow-mouth bottles, 500-ml capacity
b. Glass bottles with stopper — 300 ml capacity
c. Glass fiber filter paper — GF/C or equivalent
d. Glass funnels
e. Volumetric flasks — 1 L capacity
f. Oven — set at 103°C

4. Reagents

a. Sodium hydroxide solution, NaOH, 6 N: Refer to Section 3.5.3 and Table 3.5.
b. Aluminum chloride solution; $AlCl_3.6\ H_2O$, 6 N: Weigh 100 g $AlCl_3.\ 6\ H_2O$ and dissolve in 144 ml DDW.

Note: Aluminum chloride is highly hygroscopic and tends to form cakes, so always purchase 100-g bottles of $AlCl_3.6\ H_2O$ and dissolve the complete contents in 144 ml DDW.

c. Zinc acetate, $(CH_3\ COO)_2\ Zn.2\ H_2O$, 2 N: Weigh 220-g zinc acetate and transfer to a 1L volumetric flask. Dissolve in DDW. Make up the final volume to 1 L with DDW.
d. Hydrochloric acid, HCl, 6 N: Refer to Section 3.5.3 and Table 3.4.
e. Standard solutions
   i. Iodine solution; 0.025 N: Weigh 20 to 25 g KI and transfer to a 1L volumetric flask. Dissolve in a small amount of DDW. Add 3.2 g Iodine to the solution of KI. Dilute up to the mark with DDW after dissolution of iodine. Standardize iodine solution against 0.025 N sodium thiosulfate solution using starch as indicator.
   ii. Standard sodium thiosulfate solution; $Na_2S_2O_3.5\ H_2O$, 0.025 N: Weigh 6.025 g $Na_2S_2O_3.5H_2O$ and transfer to a 1L volumetric flask. Dissolve in DDW and make up the final volume to 1L. Standardize this solution with 0.025N potassium dichromate solution.
   iii. Potassium dichromate solution; $K_2Cr_2O_7$, 0.025 N: Keep an aliquot quantity of $K_2Cr_2O_7$ in an oven at 103°C for 2 h. Cool to room temperature in a desiccating cabinet. Weigh 1.226-g dry, cool dichromate and transfer to a 1L volumetric flask. Dissolve in DDW and dilute to 1L.
f. Sulfuric acid, H_2SO_4, 6 N: Refer to Section 3.5.3 and Table 3.4.

g. Starch solution:
   i. Measure 100 ml DDW in a 250-ml beaker and keep on a heater equipped with magnetic stirrer.
   ii. Bring to a boil.
   iii. Weigh 2-g laboratory-grade starch and suspend in a small volume of DDW.
   iv. Using magnetic stirrer, add starch suspension in small increments to boiling DDW with constant stirring. Cool the solution. Add 0.2-g salicylic acid as preservative.

5. Standardization

a. Sodium thiosulfate solution
   i. Take 2 g KI (free from iodate) in a 500-ml Erlenmeyer flask and dissolve in 100 ml DDW.
   ii. Add I ml of 6 N H_2SO_4 and 20 ml standard 0.025 N $K_2Cr_2O_7$ solution. Place the flask in the dark.
   iii. Dilute to 400 ml with DDW and titrate liberated iodine with thiosulfate.
   iv. When a pale straw color is reached, add few drops of starch indicator. A blue color will develop.
   v. Titrate with thiosulfate till the blue color disappears.
   vi. Record the reading in the burette.

$$\text{Normality of } Na_2S_2O_3 = \frac{0.025 \times 20}{\text{Volume of } Na_2S_2O_3 \text{ used, ml}}$$

Note: *20 ml of standard 0.025 N $K_2Cr_2O_7$ solution requires 20 ml $Na_2S_2O_3$ solution. If there is difference in the volume reading, calculate the exact normality of $Na_2S_2O_3$ with above-mentioned relation. If there is a significant difference in the normality of $Na_2S_2O_3$ solution, discard it. Prepare fresh solution and titrate with $K_2Cr_2O_7$ solution.*

b. Iodine solution
   i. Take 20-ml standard iodine solution in Erlenmeyer flask and add 2 ml 6 N HCl.
   ii. Titrate with standard sodium thiosulfate solution, adding few drops of starch solution indicator.
   iii. Record readings of burette.

$$\text{Normality of Iodine} = \frac{0.025 \times \text{ ml } Na_2S_2O_3 \text{ used}}{20}$$

6. Procedure

a. To a 300-ml stoppered glass bottle add 1 ml of 6 N NaOH. Fill the bottle with sample.
b. Add 1 ml of 6 N $AlCl_3$.

c. Close the bottle with stopper tightly without entrapment of air. Rotate back and forth for about one min till a floc of suspended matter is formed.

Note: The pH of the content should range from 8.0 to 9.0. This will prevent the escape of dissolved sulfide in the form of H_2S gas.

d. Allow the floc to settle down and filter the contents in a 300-ml stoppered bottle containing 2 ml of 2 N zinc acetate.

e. Add 1 ml, 6 N NaOH. Stopper the bottle without entrapment of air bubble.

f. Let the precipitate settle and filter through a glass fiber filter.

g. Take the filter paper with precipitate in a 250-ml Erlenmeyer flask and add 100 ml DDW.

h. Add 20 ml standard iodine solution to the flask. Now add 2 ml 6 N HCl.

i. Titrate this solution against standard $Na_2S_2O_3$ solution using starch as indicator.

7. Calculations

$$\text{mg Sulfide/L} = \frac{[(A \times B) - (C \times D)] \times 16000}{\text{Volume of sample, ml}}$$

Where, A = Volume of Iodine taken
B = Normality of Iodine
C = Volume of $Na_2S_2O_3$ used
D = Normality of $Na_2S_2O_3$ solution

7.5.9 Sulfite — Titrimetric Method

Source

Sulfite ions may occur in boilers and boiler feed water treated with sulfite for dissolved oxygen control. Sulfite accesses natural water through industrial wastes or effluents from treatment plants where dechlorination is carried out with SO_2.

Significance

1. Corrosion

Excessive sulfite in boiler waters is harmful because it lowers the pH and accelerates corrosion.

2. Toxicity

Water with high sulfite content should not be discharged into natural water because SO_3^{2-} is toxic to fish and other aquatic life.

3. Oxygen demand

Sulfite is an unstable component that readily oxidizes to sulfate, thus creating high O_2 demand. This affects the growth and life of aquatic animals.

Measurement

1. Principle

A water sample containing sulfite is titrated with standard KI-iodate titrant under acidic condition. Free iodine thus liberated reacts with sulfite available in the sample. The excess iodine forms a blue color with starch indicator. This is considered the end point.

Note: *As an unstable component, sulfite should be measured immediately after collection of the sample. If it is to be analyzed in the laboratory, the sample should be properly preserved before transportation (see Table 3.1).*

2. Apparatus

a. Sampling bottles — polyethylene wide-mouth bottles of 250 ml capacity
b. Oven — set at 120°C.
c. Heater with magnetic stirrer
d. Desiccating cabinet
e. Volumetric flasks — 100 ml and 1L capacity
f. Beaker — 2 L capacity
g. Erlenmeyer flask — 250 ml capacity

3. Sampling

a. Collect a fresh sample in sampling bottle avoiding contact with air.
b. Add 1 ml EDTA solution for each 100-ml sample. Do not filter. EDTA will form complexes with oxidizable materials.

4. Reagents

a. Sulfuric acid, H_2SO_4, 18 N: Refer to Section 3.5.3 and Table 3.4.
b. Standard potassium iodide-iodate solution, KI-KIO_3, 0.0125 N: Dry an adequate quantity of KIO_3 in an oven at 120°C for 4 h. Cool to room temperature in a desiccating cabinet. Weigh 0.4458 g KIO_3, 4.35 g KI and 0.310 g $NaHCO_3$ and transfer to a 1-L volumetric flask. Dissolve in DDW. Dilute this mixture to 1 L.

$$1\ L\ KI — KIO_3\ solution \equiv 500\ mg\ SO_3^{2-}$$

Note: *The titrant should be clear, not yellow in color. Any yellow color may indicate the presence of free iodine or high pH of DDW or air oxidation of KI reagent. Always store this solution in a tinted glass reagent bottle.*

c. EDTA solution: Weigh 2.5-g EDTA disodium salt and transfer to a 100-ml volumetric flask. Dissolve in DDW.

Note: *This reagent is added as a complexing agent at the time of sample collection. This helps to remove the other oxidizable materials, which interfere with the estimation of sulfite.*

d. Starch indicator: Follow the same procedure as described in Section 7.5.8.

e. Sulfamic acid, NH_2SO_3H: Use crystalline sulfamic acid.

5. Procedure

a. Add 1 ml of 18 N H_2SO_4 and 0.1 g crystalline sulfamic acid to a 250 ml Erlenmeyer flask.

Note: *Sulfamic acid is added to remove nitrite, which causes interference with sulfite estimation.*

b. Carefully add 100 ml stabilized sample into the flask avoiding contact with air. Keep the tip of pipette below the surface of the liquid.

c. Add 1 ml of starch indicator. Titrate immediately with KI-KIO_3 standard titrant with stirring.

d. Stop adding the titrant when a faint blue color develops and persists. This indicates the end point.

e. Record the reading in the burette.

f. Prepare a blank with 100 ml DDW and titrate as described for the sample.

6. Calculation

$$\text{mg sulfite/L} = \frac{(A - B) \times N \times 40000}{\text{volume of sample, ml}}$$

where A = volume of titrant required for sample, ml
B = volume of titrant required for blank, ml
N = normality of titrant solution

Since normality of titrant is 0.0125 N, the above relation is modified as follows:

$$\text{mg sulfite/L} = \frac{(A - B) \times 500}{\text{volume of sample, ml}}$$

Chapter 8

Determination of Metallic Constituents

Water can dissolve all metals to a certain extent. The concentration of metal available in water should fall under permissible limits, because excessive amounts of any metal may be health hazardous.

Metals are divided into two categories, depending on their effect on human and other organisms.

1. Nontoxic metals

Nontoxic metals are not health hazardous, but excessive amounts can impart other objectionable characteristics to water, such as taste, color and hardness. Common metals included in this category are Ca, Mg, Na, Fe, Mn, Al, Cu and Zn. Though these metals are nontoxic in nature, in large concentrations they can cause health hazards. Therefore, a permissible limit is set up for the level of these metals in the potable water supply and in water used for irrigation.

2. Toxic metals

Toxic metals are generally found in minute quantities and are harmful to humans and other organisms. The metals falling under this category are As, Cd, Pb, Cr, Hg and Ag. Of these, As, Cd, Pb and Hg are particularly hazardous. These metals are known as trace metals.

This section of the book deals with the nontoxic metals Ca, Mg, Na and Fe — the common metallic constituents of water and domestic wastewater that appears in significant concentrations.

For the estimation of toxic metals, see Ref. 10.

8.1 Calcium (Ca) — EDTA Method

For Source and Significance, refer to Section 7.4.

Measurement

1. Principle

The hardness of water is due to the combined effect of Ca^{2+} and Mg^{2+} ions present in the form of carbonates and bicarbonates in water. Both ions have a tendency to form complexes with EDTA. Mg^{2+} ions precipitate out as insoluble Mg $(OH)_2$, if the pH of water is more than 12, while Ca^{2+} ions remain in the soluble state at this pH. Therefore, at pH between 12 and 13, only Ca^{2+} ions are available in the sample to form a complex with EDTA titrant. The end point can be detected by an indicator, Eriochrome Blue Black R. This indicator is originally red in color, but changes to blue when all free Ca^{2+} ions have been complexed with EDTA.

2. Apparatus

a. Erlenmeyer conical flasks — 250 and 500 ml capacity
b. Titration assembly — refer to Section 7.2.
c. Volumetric flasks — 100 ml and 1L capacity
d. Funnel
e. Hot plate

3. Reagents

a. Hydroxide solution, NaOH or KOH, 1 N: Follow the procedure described in Section 3.5.3 and Table 3.5.
b. EDTA Titrant, ethylene diamine tetra-acetic acid-disodium salt, $Na_2H_2C_{10}O_8N_2 \cdot 2H_2O$, 0.01 M: Prepare in the manner described in Section 7.4.

$$1 \text{ ml } 0.01 \text{ M EDTA solution} \equiv 400.8 \ \mu g \ Ca^{2+}/ml$$

c. Indicator solution — Eriochrome Blue Black R, Sodium-1-(2-hydroxyl-1 naphthylazo)-2-naphthol-4-sulfonic acid: Weigh 0.4 g of dye and transfer to a 100-ml volumetric flask. Dissolve in methyl alcohol. Make up the final volume to 100 ml with alcohol. This indicator solution can also be purchased from any international chemical manufacturing company.
d. Ammonium hydroxide, NH_4OH, 24% (v/v): Refer to Section 7.4.
e. Hydrochloric acid, HCl, 50% (v/v): Refer to Section 7.4.

f. Standard calcium solution, $CaCO_3$ anhydrous: Follow the procedure of preparation described in Section 7.4.

$$1.0 \text{ ml standard calcium solution} \equiv 1 \text{ mg } CaCO_2$$

4. Standardization

Prepare the calibration standards containing 10, 20 and 50 mg/L $CaCO_3$ and standardize as follows:

a. Into three different 100-ml volumetric flasks measure 1.0, 2.0 and 5.0 ml standard calcium solution.
b. Dilute the solution in each flask up to the mark with DDW.
c. Transfer these standards into three different 250-ml Erlenmeyer flasks.
d. Add required volume of 1 N NaOH solution to bring the pH of standards between 12 and 13. Record this volume.
e. Add 2–3 drops of Eriochrome indicator to each flask. A red color will be developed at this stage.
f. Titrate with EDTA titrant with constant stirring. Stop adding the titrant when red color changes to blue.
g. Wait for 1 min; if red color appears again, titrate till blue color is obtained. This is the end point.
h. These standard solutions should utilize approximately 1, 2 and 5 ml of the titrant.
i. Set up a reagent blank with DDW instead of standard with all reagents and titrate in the same way as standard.
j. Calculate mg $CaCO_3$ equivalent by following relation:

$$D = \frac{\text{mg } CaCO_3}{(A - B)} = 1$$

where A = volume of titrant used for standard
B = volume of titrant used for Blank
D = mg $CaCO_3$ equivalent to 1.00 ml EDTA titrant

5. Procedure

a. Take 100-ml sample or aliquot quantity of sample diluted to 100 ml in a 250-ml Erlenmeyer flask.
b. Add the same volume of 1N NaOH solution used during standardization to acquire a pH level between 12 and 13 in the sample. Add 2–3 drops of Eriochrome indicator also.
c. Titrate with 0.01 M EDTA titrant slowly, stirring constantly, until the color changes from red to blue. Follow the same procedure as described for standardization to obtain the end point.

6. Calculations

$$\text{Concentration of } Ca^{2+} \text{ (mg/L)} = \frac{(C - B) \times D \times 400.8}{\text{volume of sample, ml}}$$

where B = volume of titrant consumed for Blank, ml
C = volume of titrant consumed for Sample, ml

7. Important Instructions

a. Alkalinity more than 300 mg/L may cause interference in the detection of end point.
b. In this procedure, alkali is added to achieve very high pH (12–13), so the titration should be performed immediately after the addition of alkali, i.e., within 5 min.
c. The sample of wastewater or contaminated water should be pretreated as described in Section 7.4.

8.2 Magnesium (Mg) — Gravimetric Method — Detectable Level: ≤ 60 mg/L Mg

Magnesium is the metal concerned with the hardness of water and is a common constituent of natural water.

For Source and Significance, see Section 7.4.

Measurement

1. Principle

Mg^{2+} ions together with Ca^{2+} ions are responsible for the hardness of water. Di-ammonium hydrogen phosphate used in this procedure quantitatively precipitates magnesium in ammonical solution as magnesium ammonium phosphate. The concentration of Mg is determined gravimetrically by igniting the precipitate of magnesium ammonium phosphate at 550°C, which will be converted to magnesium pyrophosphate. The weight of magnesium pyrophosphate is recorded.

Note: The procedure involves the removal of ammonium salts and oxalates before precipitation of Mg. It is applied to the filtrate and washings obtained after precipitating calcium with ammonium oxalate.

2., Apparatus

a. Volumetric flasks of different capacities
b. Erlenmeyer flasks of different capacities
c. Magnetic stirrer
d. Heating mantle — adjusted at 90°C

e. Crucible

f. Muffle furnace – set at 550°C

3. Reagents

a, Methyl red indicator solution: Weigh 0.1-g methyl red sodium salt and transfer to a 100-ml volumetric flask. Dissolve in DDW and dilute to 100-ml with DDW.

b. Hydrochloric acid, HCl:

i. 50% (v/v) concentration: Carefully add 500 ml concentrated HCl to 500-ml DDW.

ii. 10% (v/v) concentration: Dilute 100 ml concentrated HCl to 1L with DDW.

iii. 1-% (v/v) concentration: Dilute 10 ml concentrated HCl to 1L with DDW.

c. Ammonium oxalate solution, $(NH_4)_2C_2O_4 \cdot H_2O$: Weigh 4.0 g $(NH_4)_2 C_2O_4 \cdot H_2O$ and transfer to a 100-ml volumetric flask. Dissolve in DDW and make up the final volume to 100 ml. If any turbidity appears, filter the solution.

d. Ammonium hydroxide, NH_4OH:

i. Concentrated

ii. 24% (v/v): Take 240 ml concentrated NH_4OH in 1-L volumetric flask and dilute with DDW to 1 L.

iii. 5% (v/v): Dilute 12.5 ml concentrated NH_4OH to 500 ml with DDW.

iv. 1% (v/v): Dilute 1 ml concentrated NH_4OH to 100 ml with DDW.

d. Nitric acid, HNO_3: Concentrated

e. Diammonium hydrogen phosphate solution, $(NH_4)_2 HPO_4$: Weigh 30.0 g $(NH_4)_2 HPO_4$ and transfer to a 100-ml volumetric flask. Dissolve in DDW. Bring the final volume up to the mark with DDW.

3. Procedure

a. Removal of calcium from the sample:

i. Take 250-ml sample or an aliquot diluted to 250 ml in a 500-ml Erlenmeyer flask and add 2–3 drops of methyl red indicator.

ii. Neutralize with 50% (v/v) HCl. Keep the flask on a hot plate for boiling. Boil for 1 min.

iii. Remove flask from hot plate and add 50 ml ammonium oxalate solution.

Note: If any precipitate appears in Step III, add 50% (v/v) HCl in excess to re-dissolve it.

iv. Now keep the flask for heating on a mantle adjusted, at 90°C. At this stage, the solution should not boil.

v. Turn on the magnetic stirrer and slowly add 24% (v/v) NH_4OH solution dropwise from a pipette.

vi. Continue addition of NH_4OH, till a highly turbid solution is obtained.

vii. Now digest this sample for 1 h at 90°C.

viii. Filter the digested sample through GF/C filter under vacuum.

ix. Wash the flask and filter twice with small volume of 1% (v/v) NH_4OH solution.

x. Combine the filtrate and washings in an Erlenmeyer flask to determine magnesium.

b. Determination of magnesium: dual precipitation method

i. Add 0.5–1.0 ml of methyl red solution to the combined filtrate and washings obtained from Step a.

ii. Bring the volume to 150 ml with DDW and acidify with 5 ml of 50% (v/v) HCl.

iii. Add 10-ml $(NH_4)_2$ HPO_4 solution. Cool the solution.

iv. Keep the flask on magnetic stirrer and add concentrated NH_4OH dropwise with constant stirring until the color of the solution changes to yellow.

v. Stir for 5 min more. Then add 5 ml concentrated NH_4OH and stir vigorously for 10 min more. At this stage, magnesium will be precipitated as magnesium ammonium phosphate.

vi. Let the solution stand overnight at room temperature. Filter to remove precipitate.

vii. Wash the filter paper with small volume of 5% (v/v) NH_4OH solution.

viii. Discard the filtrate and washings.

ix. Dissolve the precipitate thus collected on filter paper in warm 10% (v/v) HCl solution.

x. Wash filter paper with hot 1-% (v/v) HCl solution.

xi. Repeat Steps I to VII for re-precipitating magnesium.

xii. In the meantime, prepare a crucible for gravimetric determination. Keep the crucible for 2 h in a muffle furnace at 550°C. Cool it by keeping in a desiccating cabinet. Weigh to a constant weight. Record the weight.

xiii. After filtration and washing, transfer the filter paper to pre-weighed crucible.

xiv. Dry filter paper along with precipitate and ignite at 550°C in a muffle furnace.

xv. Continue ignition until the residue becomes white.

xvi. Further ignite for 30 min. at 1,100°C. Magnesium is obtained as magnesium pyrophosphate.

xvii. Cool the crucible by keeping it in a desiccating cabinet and weigh after cooling.

xviii. Repeat the procedure of desiccating, cooling and weighing until a constant weight is achieved.

4. Calculation

$$\text{Concentration of } Mg^{2+}\text{, mg/L} = \frac{(A - B) \times 218.5}{\text{volume of sample, ml}}$$

where A = weight of $Mg_2P_2O_7$ (magnesium pyrophosphate) residue
B = weight of crucible

8.3 Iron (Fe) — Phenanthroline Method — Detectable Range: 0.02 to 3.0 mg/L Fe

Measurement

1. Principle

Total iron available in the sample is converted to the reduced ferrous (Fe^{2+}) state by boiling the sample with HCl and hydroxylamine. Iron present in the ferrous state reacts with 1, 10-phenanthroline under acidic environment (pH 3.0 to 3.5) and produces an orange–red complex. The absorbance of color is read at 515-nm wavelength. The intensity of the color is proportional to the concentration of iron available in water or wastewater sample.

2. Apparatus

a. UV spectrophotometer — set at 515 nm wavelength
b. Homogenizer
c. Heating mantle
d. Volumetric flasks with different capacities
e. Erlenmeyer flasks with different capacities

3. Reagents

a. Iron-free water: Use DDW to prepare standards, solutions and all reagents.
b. Hydrochloric acid, HCl: concentrated
c. Hydroxylamine solution, NH_2OH. HCl, 10% solution: Weigh 10 g NH_2OH. HCl and transfer to a 100-ml volumetric flask. Dissolve in DDW. Make up the volume to 100 ml with DDW.
d. Ammonium acetate buffer:
   i. Weigh 250 g ammonium acetate and transfer in a 1-L volumetric flask. Dissolve in DDW. Add DDW to bring volume to 300 ml.
   ii. Add 700 ml concentrated acetic acid.

Note: *With every buffer preparation, prepare new reagents and standard solutions because ammonium acetate contains significant amount of iron.*

e. Sodium acetate solution, CH_3COONa. 3 H_2O: Weigh 100-g CH_3COONa. 3 H_2O and transfer to a 500-ml volumetric flask. Dissolve in DDW. Bring the final volume up to the mark with DDW.
f. Phenanthroline solution, $C_{12}H_8N_2$. H_2O: Weigh 0.1 g 1, 10-phenanthroline monohydrate and transfer to a 100-ml volumetric flask. Dissolve with constant stirring in DDW acidified with 1 ml of concentrated HCl. Bring the final volume to 100 ml.

g. Sulfuric acid, H_2SO_4, 40% (v/v): Carefully add 40 ml concentrated H_2SO_4 to 60 ml DDW. Cool the solution to room temperature.

h. Potassium permanganate, $KMnO_4$, 0.2%: Weigh 0.2 g $KMnO_4$ and transfer to a 100-ml volumetric flask. Dissolve in DDW. Bring the final volume up to the mark with DDW.

i. Iron stock solution:

   i. Weigh 0.7022 g ferrous ammonium sulfate hexahydrate, $Fe\ (NH_4)_2\ (SO_4)_2 \cdot 6\ H_2O$ and transfer to a 1-L volumetric flask. Dissolve in minimum volume of DDW.

   ii. Add 5 ml 40% H_2SO_4.

   iii. Add 0.2% $KMnO_4$ solution drop by drop until a faint pink color persists in the solution after stirring well.

   iv. Dilute up to the calibration mark with DDW. Stopper and invert the flask several times to mix thoroughly.

$$1.0 \text{ ml stock solution} \equiv 0.1 \text{ mg } Fe^{2+} \text{ ion}$$

Note: *Since this solution is very unstable, always prepare fresh stock solution before analysis.*

h. Iron standard solution: Measure 10 ml of iron stock solution in a 100-ml volumetric flask and dilute with DDW up to the mark.

$$1.0 \text{ ml standard solution} \equiv 0.01 \text{ mg } Fe^{2+} \text{ ion}$$

Note: *Always prepare fresh standard solution before analysis.*

4. Standardization

Prepare calibration standards containing 0.1, 0.5, 1.0, 2.0 and 3.0 mg/L Fe^{2+} ion and standardize as follows:

a. Into five different 100-ml volumetric flasks measure 1, 5, 10, 20 and 30 ml of iron standard solution.

b. Add 50 ml DDW in each flask with a graduated cylinder.

c. Add 10-ml ammonium acetate buffer and 5 ml 1, 10-phenanthroline in each flask.

d. Dilute solution in each flask up to the mark with DDW. Stopper and invert each several times to mix.

e. Wait for about 15 min. for color development. An orange color will appear.

f. Prepare a blank with DDW instead of standard solution.

g. Read the absorbance of each standard against the blank.

h. Plot a calibration curve of absorbance against the concentration of Fe^{2+} ions in mg/L.

5. Procedure

a. Digestion of sample for total iron
   i. Homogenize the sample and measure 100 ml in a 500-ml Erlenmeyer flask.
   ii. Add 5 ml concentrated HCl and 2 ml hydroxylamine solution to the sample.
   iii. Add a few glass beads to the flask and keep it on a heating mantle in a fume hood.
   iv. Allow the solution to boil until the volume reduces to 15 to 20 ml.
   v. Cool the solution to room temperature.

b. Estimation
   i. Transfer the digested sample to a 100-ml volumetric flask.
   ii. Add 50 ml DDW to the flask.
   iii. Develop the color with 1, 10-phenanthroline as described in Step 4.
   iv. Read the absorbance of orange color at 510 nm.

6. Calculation

Read the concentration of Fe in the sample directly from the calibration curve.

8.4 Metal Detection by Atomic Absorption (AA) — Spectrophotometric Method — Sample Preparation

Description

The term "metals" represents all metals present both in inorganic and organic compounds in dissolved and suspended states in water and wastewater samples. The following terms are related with this estimation:

1. Dissolved Metals: Metals that pass through a 0.45 μm membrane filter
2. Suspended Metals: Metals that retained on a 0.45 μm membrane filter
3. Total Metals: Total metal content includes both dissolved and suspended metal components.

This section describes the procedure used for digestion of samples to detect total metals. Samples are taken directly without filtration for digestion.

Measurement

1. Principle

For the estimation of total metals, the sample is taken directly without filtration for digestion with nitric acid. The digestion process converts the metals present in combined state into free state. The metal content in the digested sample is measured by atomic absorption spectroscopy.

2. Apparatus

a, Sampling bottles – polyethylene

b. Volumetric flasks – 1L capacity

c. Beaker – 250 ml capacity

d. Watch glass

e. Hot plate

3. Reagents

a. Nitric acid, HNO_3:

i. Concentrated

ii. 50% (v/v) solution: Measure 500 ml DDW in a 1L volumetric flask. Add 500 ml concentrated HNO_3 to it with full precaution. Mix it carefully and cool.

b. Hydrochloric acid, HCl, 50% (v/v): Measure 500 ml DDW in a 1L volumetric flask. Add 500 ml concentrated HCl to it and mix well.

c. Sodium hydroxide, NaOH, 5N: See Section 3.5.3 and Table 3.5.

4. Procedure

a. Sampling:

i. Use clean polyethylene bottles for sampling. Rinse with 50% (v/v) HNO_3 and finally rinse with DDW.

ii. Preserve the sample by bringing its pH to 2.0 with 50% (v/v) HNO_3.

b. Digestion – Total metals

i. Transfer 100 ml (if expected metal concentration is 1 mg/L) or take an appropriate volume diluted to 100 ml into a beaker and add 5 ml concentrated HNO_3.

ii. Place the beaker on a hot plate and evaporate the solution to the lowest volume between 5–10 ml.

Note: The sample must not boil during this process.

iii. Cool the beaker and add another 5 ml of the concentrated HNO_3.

iv. Cover the beaker with a watch glass and return it to the hot plate. Adjust the temperature of hot plate in such a way that a gentle refluxing can occur without boiling the solution. If required, add additional acid. The completion of digestion is indicated by the appearance of light-colored, clear solution.

v. Again evaporate the solution to lowest volume and cool the beaker. If any residue or precipitate appears after evaporation, add 5 ml of 50% (v/v) HCl solution.

vi. Wash down the beaker walls and watch glass with minimum volume of DDW. Warm the beaker. Add 5 ml of 5.0 N NaOH and quantitatively transfer the sample to a 100-ml volumetric flask. Also transfer the washings to the flask.

vii. Examine the pH of the sample. If it is lower than 4.0, adjust it with dropwise addition of 5.0 N NaOH solution. Mix thoroughly and examine the pH after each addition. Adjust the pH to 4.0. Make up the volume to 100 ml with DDW.

viii. A reagent blank should also be digested as described for sample.

ix. Multiply the result by the correction factor as presented in Table 8.1.

TABLE 8.1
Volume of Sample Taken for Digestion in the Analysis of Metals

Expected metal concentration, mg/L	Sample vol. for digestion, ml	Vol. of 50% (v/v) HCl, ml	Final volume, ml	Correction factor
1	50	10	200	4
10	5	10	200	40
100	1	25	500	500

8.5 Sodium (Na) – Atomic Absorption Spectrophotometric Method – Detectable Range: 0.1 to 3.0 mg/L Na

Source

Na is the most common nontoxic metal found in natural waters. It is abundant in the earth's crust. Sodium salts are highly soluble in water, so are leached from soil and rocks. Sodium salts are commonly used in various industries, thus are present in significant quantities in industrial wastes.

Significance

Excessive concentrations of Na impart a bitter taste to drinking water. It is hazardous for people suffering from cardiac and kidney ailments. High concentrations of Na in public water supply and treated effluent used for irrigation deteriorates the physical condition of soil by reducing its permeability. This affects the growth and yield of plants. Sodium is corrosive to metal surfaces too.

Measurement

1. Principle

Sodium in water or wastewater sample is detected by Flame Atomic Absorption Spectroscopy. When a sample containing metallic salt is aspirated into a flame (produced by burning of acetylene in air) it vaporizes. The vapors thus produced contain atoms of metal both in excited and ground (unexcited) states.

The free unexcited atoms in the flame are in much greater abundance than excited atoms. When a light of resonance wavelength is passed through the flame, these atoms absorb radiant energy of specific resonance wavelength. The amount of light absorbed will be proportional to the number of metallic atoms present in the flame. This is the basic principle of Atomic Absorption Spectroscopy (AAS). Sodium is detected by AAS at 589.6 nm.

2. Important safety measures

The following safety measures should be taken before working on an atomic absorption spectrophotometer (AAS):

a. The instrument must be installed in a well-ventilated laboratory equipped with adequate exhaust system.
b. It is essential to read the operation manual and the manufacturer's instructions before handling AAS.
c. Gas cylinders must be kept in special safety cabinets with proper ventilation. For adequate handling and storage of such cylinders, follow the instructions in Section 2.2.2.a.
d. Make periodic checks for gas leakage by applying soap solution to joints and seals.
e. When the instrument is switched off, close the fuel gas supply valve tightly and bleed the gas line to the atmosphere through the exhaust vent.
f. The gas-supplying pipes must be securely fixed on the walls.
g. A nitrous oxide cylinder must not be used after the pressure has dropped to 100 psi.
h. Never view the flame or hollow cathode lamp directly. Always wear protective safety goggles to protect the eyes from exposure to U V light.
i. Never leave the instrument unattended after igniting the flame.

Detailed safety rules and precautions are cited in Reference 27.

3. Apparatus

a. Atomic absorption spectrophotometer: Any of the latest commercially available models are suitable.

Caution: *Read the instructions of the manufacturer and follow them strictly to operate the instrument.*

b. Fuel – acetylene:

Standard commercially available acetylene is used as fuel. Minimum acceptable purity of acetylene is 99.5% by volume.

Impurity: Acetone is a common contaminant of acetylene. The presence of acetone will produce a flame with a slight pink tinge with white flakes and sparks. This will yield abnormally high pulsating background signals and interfere with analytical estimation of metals. Hence, the following precautions should be taken while using acetylene as fuel:

i. Always use acetylene gas at a pressure of 15 psi (103 k Nm^{-2}). Never run acetylene gas at pressure higher than this level, as the cylinder might explode.

ii. A full acetylene cylinder usually shows the pressure of 250 psi at 20°C. Never use the cylinder if pressure drops to 50 psi; this will allow the access of acetone into the system. Acetone can damage the plastics in the piping system of the instrument.

iii. Never use acetylene in an extremely cold or hot (above 48°C) environment.

iv. Prevent contact between the gaseous acetylene and silver, mercury or chlorine.

v. Do not use copper tubing for the gas-supplying system. Use tubing made up of brass containing less than 65% Cu or any other material that does not react with gaseous acetylene, e.g., galvanized iron.

c. Oxidant – air:

Air is supplied for the burning of acetylene to get a uniform and stable flame. Air should be dry and free from oil and other foreign substances. This is ensured by passing air through a good-quality filter installed on the air supply line at the point of use. Air compressors having a filter-regulator assembly are generally used for this purpose.

d. Hollow cathode lamps:

Sodium hollow cathode lamp is used for estimation of sodium.

e. Pressure reducing valves:

To maintain proper supply of fuel and oxidant at required pressures, use pressure-reducing valves. Use a separate valve for each gas.

f. Vent:

During operation of AAS the burner produces a large quantity of hot exhaust gases. In addition, certain elements or materials produce noxious / toxic by-products. Hence, a proper vent at about 15 to 30 cm above the burner is essential.

Caution: Use silver solder or rivets for construction of venting system. Do not use soft (lead, tin) solder.

g. Drainage:

The spectrophotometer premix chamber is connected to a drain trap to collect unused (waste) sample from the premix chamber.

Caution: Always ensure that the drain line is connected to the premix chamber and drain trap before performing flame operation.

h. Oven – set at 105°C

i., Desiccating cabinet

4. Reagents

a. Acetylene: see 2 b.

b. Air: see 2c.

c. Metal-free water: Use DDW for preparing standard solutions, reagents and for dilutions.

d. Sodium stock solution, sodium chloride, NaCl:

i. Dry an adequate quantity of NaCl in an oven at 105°C for 2 h. Bring to room temperature in a desiccating cabinet.
ii. Weigh 2.542 g dried NaCl and transfer to a 1-L volumetric flask. Dissolve in DDW. Bring to the final volume with DDW.

$$1.0 \text{ ml stock solution} \equiv 1.0 \text{ mg Na}$$

e. Sodium standard solution: Measure 10 ml of sodium stock solution in a 100-ml volumetric flask and dilute up to the mark with DDW.

$$1.0 \text{ ml standard solution} \equiv 0.1 \text{ mg Na}$$

5. *Standardization*

Prepare the calibration standards containing 0.1, 1.0, 2.0 and 3.0-mg/L Na concentration and standardize by following method:

a. Into four different 100-ml volumetric flasks measure 0.1, 1.0, 2.0 and 3.0 ml of sodium standard solution.
b. Bring up the volume to 100 ml with DDW in each flask.
c. Prepare blank with 100 ml DDW containing 0.2 ml concentrated HNO_3.
d. Aspirate the blank and the standards following the instructions of the manufacturer.
e. Record the absorbance in each case.
f. Plot the standard curve of absorbance against the concentration of Na in mg/L units.

Note: *If the instrument is equipped with a direct concentration readout do not prepare the calibration curve.*

6. *Procedure*

a. Rinse the nebulizer with blank before analyzing the samples.

Note: *The function of the nebulizer is to produce a mist or aerosol of the sample.*

b. Aspirate each sample and follow the same procedure as described in Step 3. Record the absorbance.
c. Aspirate the blank between sample analyses.

7. *Calculation*

a. Calculate the concentration of Na in the sample by the calibration curve or by direct readout.
b. If the sample is diluted, then:

$$\text{Na, mg/L} = \text{absorbance} \times \text{df}$$

where df is the dilution factor
If 1 ml sample is diluted to 10 ml then df = 10.
If 1 ml sample is diluted to 100 ml then df = 100 and so on.

8.6 Potassium (K) – Atomic Absorption Spectrophotometric Method – Detectable Range: 0.2 to 4.0 mg/L K

Measurement

1. Principle

Potassium (K) in water and wastewater samples is determined by flame AA spectroscopy.

2. Apparatus

Same as described in Section 8.5 except here, a potassium hollow cathode lamp is used for detection.

3. Reagents

a. Acetylene: Refer to Section 8.5.
b. Air: Refer to Section 8.5.
c. Metal-free water: Use DDW.
d. Potassium stock solution, potassium chloride, KCl:
   i. Dry an aliquot quantity of KCl in an oven set at 105°C for 2h. Cool to room temperature in a desiccating cabinet.
   ii. Weigh 1.907 g dry and cool KCl and transfer to a 1-L volumetric flask. Dissolve in DDW. Bring the final volume to 1 L with DDW.

$$1.0 \text{ ml stock solution} \equiv 1.0 \text{ mg K}$$

e. Potassium standard solution: Measure 10 ml of potassium stock solution in a 100-ml volumetric flask and dilute up to the mark with DDW.

$$1.0 \text{ ml standard solution} \equiv 0.1 \text{ mg K}$$

4. Standardization

Prepare the calibration standards containing 0.2, 1.0, 2.0 and 4.0 mg K/L and standardize by the following method:

a. Into four different 100-ml volumetric flasks measure 0.2, 1.0, 2.0 and 4.0 ml of potassium standard solution.

b. Follow the same procedure of standardization and estimation as described in Section 8.5 for the estimation of sodium.

5. Calculation

Follow the same procedure of calculation as described in Section 8.5.

Chapter 9

Determination of Nutrients

Nutrients are the elements essential for the growth, propagation and activity of plants, animals, even aquatic species, and microbes. A number of elements fall under this category, but the major elements are carbon, nitrogen and phosphorus. Carbon is readily available from many sources. The major source is CO_2. This is available in abundance in the environment. Carbon dioxide is the main product of decaying organic matter, respiration of living beings and of industries having combustion processes.

Among nutrients, nitrogen and phosphorus are considered the limiting factors for the growth of plants and animals. These essential elements are described in detail in the following sections.

9.1 Nitrogen

Source

Nitrogen is the most abundant element in the environment (about 79% by weight, as molecular N_2). Nitrogen is the major constituent of organic molecules such as amino acids, the building blocks of proteins in animal and plant cells. Nitrogen is the constituent of another important organic molecule, Nucleic acid. Nucleic acids are polynucleotides, which are responsible for reproduction. Nucleotides are the vital components to control metabolic processes and transportation of chemical energy in the living systems. Thus, dead and decaying animals and plants also contribute a large amount of nitrogen in various forms to the environment. Animal wastes, sewage, industrial effluents and agricultural wastes are some other sources of nitrogen. These nitrogen-rich wastes can be discharged directly into streams or can enter waterways through surface run-off or ground-water discharges.

Chemistry

In the environment, nitrogen undergoes a series of redox reactions depending on the presence (aerobic) and absence of oxygen (anaerobic). Aerobic and anaerobic processes convert nitrogen into various derivatives such as ammonia, nitrites and nitrates.

The process of nitrogen assimilation in the environment comprises the following three major steps:

1. Conversion of nitrogen to various derivatives

The conversion of nitrogen to ammonia, nitrites or nitrates can take place by following methods:

a. Nitrogen fixation: Nitrogen fixation refers to the conversion of environmental nitrogen to plant protein through reduction of nitrogen to ammonia. It is carried out by special nitrogen fixing bacteria, e.g., *Azobacter* or *Anabaena*. These microbes can fix atmospheric N_2 in the roots of leguminous plants like peas and beans.

b. Atmospheric nitrogen to nitrates: By the action of electric discharge and cosmic radiation, the atmospheric N_2 is oxidized to N_2O_5 (nitrogen oxide). This oxide on hydrolysis produces Nitric acid, HNO_3 which is carried to the earth with rainfall. Nitrates are also produced in some commercial processes such as the manufacture of fertilizers.

c. Nitrogen-containing chemicals to plant protein: Ammonia and ammonium compounds (mainly urea) are applied to the soil to produce plant proteins.

d. Animal proteins: Animals and human beings are incapable of utilizing atmospheric N_2. Hence, they depend on plants and other animals to provide nitrogen in the form of proteins. This protein undergoes a series of reactions during metabolism and releases urea, a nitrogen-containing waste found in feces and urine. Urea on hydrolysis produces ammonia in the wastes. Hence, a considerable amount of ammonia is present in wastewater.

2. Nitrification

Ammonia present in water and wastewater undergoes oxidation by aerobic autotrophic microorganisms in two steps:

a. Conversion of ammonia to nitrites: The aerobic microbes *Nitrosomonas* derive energy from oxidation of ammonia to nitrites as shown in Equation 9.1. This energy is utilized for the formation of new bacterial cells and for their activity.

$$\begin{array}{c} 5\ CO_2 + 55\ NH_4^+ + 76\ O_2 \\ \downarrow \text{Nitrosomonas} \\ \underset{\text{(Nitrosomonas new cells)}}{\underset{\downarrow}{C_5H_7O_2N}} + 54\ NO_2 + 52\ H_2O + 109\ H^+ \end{array} \tag{9.1}$$

b. Oxidation of nitrites to nitrates: Nitrobacter is another species of aerobic bacteria that oxidizes nitrites to nitrates to obtain energy for their growth and vital functions.

$$\begin{array}{c} 5\ CO_2 + NH_3 + 195\ O_2 + 2\ H_2O + 400\ NO_2^- \\ \downarrow \text{Nitrobacter} \\ C_5H_7O_2N + 400\ NO_3^- \\ \downarrow \\ \text{(Nitrobacter new cells)} \end{array} \tag{9.2}$$

Together, these two oxidation steps describe the nitrification process. According to stoichiometry of these reactions *nitrosomonas* requires 3.22 g molecular O_2 to oxidize 1 g ammonical N and *nitrobacter* requires 1.11 g O_2 to oxidize 1 g NO_2^- to NO_3^-.

These are basic nitrification reactions taking place in activated sludge during treatment of wastewater.

3. Denitrification

This is the last step in the nitrogen cycle, in which nitrites and nitrates are reduced to molecular N_2 under anaerobic conditions. This reaction is carried out by denitrifying bacteria e.g. *Pseudomonas denitrificans*. This microbe utilizes NO_3 as a source of energy.

Thus, atmospheric nitrogen is recycled to the environment as molecular N_2 after passing through these microbiological transformation reactions. The transformation of atmospheric N_2 to different derivatives, animal and plant proteins and finally converting back to molecular N_2 is known as the nitrogen cycle. It is diagrammed in Fig. 9.1.

Significance

The estimation of N as total nitrogen comprising $NH_3 - N$, $NO_2 - N$, $NO_3 - N$ and organic N in water and wastewater are of great significance because of following reasons:

1. Biological treatment

All biological treatment processes depend on the growth and oxidation ability of the microbial ecosystem. Nitrogen is the essential growth factor to form new active microbial cells. Hence, this estimation provides knowledge of total nitrogen content available in water and wastewater for microbial propagation and activity. This helps personnel in designing, operating, and monitoring the activated sludge process in wastewater-treatment plants.

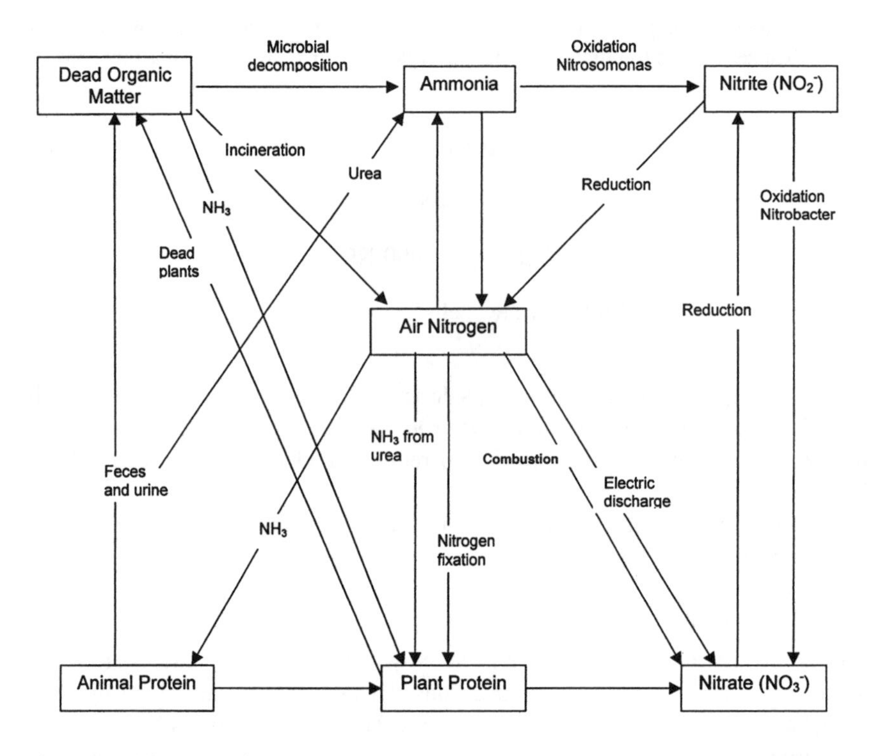

Figure 9.1 - Nitrogen cycle

2. *Eutrophication*

High nitrogen content in water streams may cause eutrophication, i.e., the excessive growth of algae. The death and subsequent decomposition of algal growth produces an oxygen demand that depletes the level of dissolved oxygen in water and threatens aquatic life.

3. *Indicator of water quality*

Nitrogen content serves as an indicator of water quality as well as the extent of water pollution. If a water stream contains a large quantity of ammonia, that water is contaminated with domestic or industrial wastes — or both. An over abundance of ammonia points out the recent discharge of waste to the water stream. If water contains mostly nitrates, this means it was polluted a long time ago, because fresh sewage contains ammonia in significant concentration. As time progresses, NH_3 oxidizes to nitrites and nitrates at the expense of the O_2 dissolved in the water. Thus, the level and period of contamination can be assessed with estimation of nitrogen content.

4. *Nitrate poisoning*

An excessive amount of nitrates in potable water can cause serious health hazards, even death, in infants. The lower acidity in an infant's intestinal tract permits the growth of nitrate reducing bacteria which converts NO_3 to NO_2. Nitrites are then absorbed into the blood stream. Nitrites have a greater affinity for hemoglobin than does O_2. Hence, it replaces O_2 in the blood, causing eventual suffocation. This results in the bluish coloration of the human body that is called methemoglobinemia.[28]

5. *Anaerobiosis in treatment*

Prolonged detention of activated sludge in treatment process results in the formation of anaerobic pockets in settling tanks. These are the best sites for reduction of nitrates to N_2 gas. Formation of gaseous nitrogen produces buoyancy and reduces the settling quality of sludge. This condition is known as rising sludge.[29] This affects the quality of secondary effluent and interferes with the further treatment of wastewater. The accumulation of scum may generate a foul smell in treatment plants and their vicinity.

Because of these factors the estimations of total N and various derivatives are highly essential. In this section, the estimation procedure of each nitrogen derivative is discussed in detail.

9.2 Ammonia – Titrimetric Method – Detectable Limit: >5 mg/L NH_3 – N

For Source and Significance, see Section 9.1.

Measurement

1. *Principle*

The sample containing ammonia is buffered at pH 9.5 with a borate buffer and distilled in a Kjeldahl distillation unit. The distillate is collected in the boric-acid-solution-containing indicator. The level of NH_3 is estimated by titration with 0.02 N H_2SO_4 titrant. The occurrence of a pale lavender color indicates the end point.

Note: The sample is buffered to control the hydrolysis of cyanates and other nitrogen-containing organic compounds.

2. *Apparatus*

a. Kjeldahl distillation unit — This unit consists of a Kjeldahl flask of 800-ml capacity connected to a condenser through an adapter. The tip of the condenser is attached to a delivery tube that reaches to the bottom of the receiver, which collects the distillate. The unit is shown in Fig.9.2.

b. Oven – set at 103°C

c. Desiccating cabinet

d. Volumetric flasks — of different capacities

e. Measuring cylinders — of different capacities

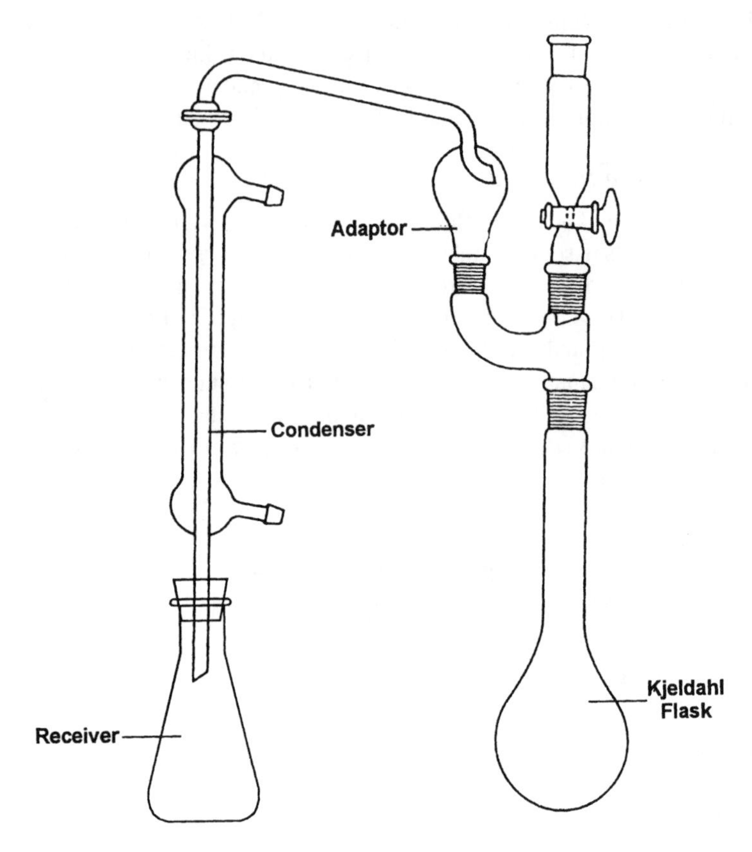

Figure 9.2 - Kjeldahl distillation unit

3. Reagents

Prepare all reagents in NH_3 -free water.

a. Ammonia-free water (AFW): Single distilled water is usually not free from ammonia. Hence, DDW should always be used. For the preparation of DDW, see Section 3.3.3.

b. Sodium hydroxide, NaOH; 0.1 N: Refer to Section 3.5.3 and Table 3.5.

c. Borate buffer solution: Weigh 4.95 g sodium tetraborate ($Na_2B_4O_7$) or 9.45 g $Na_2B_4O_7.10\ H_2O$ and transfer to a 1-L volumetric flask. Dissolve in fresh AFW. Add 88-ml of 0.1 N NaOH solution. Dilute up to the mark with AFW.

d. Sodium hydroxide, NaOH; 6.0 N: Refer to Section 3.5.3 and Table 3.5.

e. Dechlorinating agent, Sodium thiosulfate, $Na_2S_2O_3.5\ H_2O$; 0.014 N: Weigh 0.35 g $Na_2S_2O_3.5\ H_2O$ and transfer to a 100-ml volumetric flask. Dissolve in fresh AFW and dilute to 100 ml.

f. Neutralizing solutions:

   i. Sulfuric acid, H_2SO_4; 1 N: Refer to Section 3.5.3 and Table 3.4.

   ii. Sodium hydroxide, 1 N: Refer to Section 3.5.3 and Table 3.5.

g. Indicating-boric-acid solution:

i. Mixed indicator: Solution A — Weigh 0.3 g methyl red indicator and transfer to a 250-ml volumetric flask. Dissolve in 150 ml 95% ethyl alcohol.

Solution B – In another 100-ml volumetric flask weigh 0.15 g methylene blue and dissolve in 100 ml 95% ethyl alcohol.

Add solution B to solution A. Mix thoroughly to make 250-ml solution of uniform consistency.

ii. Indicating-boric-acid solution: Weigh 20 g boric acid, H_3BO_3, and transfer to a 1-L volumetric flask. Dissolve in fresh AFW. Add 10 ml mixed indicator solution in it. Bring the final volume to 1 L with AFW.

Note: Prepare fresh solution every month.

h. Standard H_2SO_4 titrant, 0.02 N: Prepare and standardize 0.02 N H_2SO_4 standard solution as described in Section 7.2.

$$1 \text{ ml } 0.02 \text{ N } H_2SO_4 \equiv 280 \text{ } \mu g \text{ N}$$

i. Stock ammonium chloride, NH_4Cl solution:

i. Keep an adequate quantity of ammonium chloride in an oven at 103°C for 1 h. Cool to room temperature in a desiccating cabinet.

ii. Weigh accurately 3.819 g NH_4Cl and transfer to a 1-L volumetric flask. Dissolve in fresh AFW. Dilute with AFW up to the mark.

$$1 \text{ ml stock } NH_4Cl \text{ solution} \equiv 1.0 \text{ mg } NH_3 - N$$

Note: This solution is stable for about six months.

j. Standard ammonium chloride, NH_4Cl solution: Measure 5 ml of stock NH_4Cl solution in a 500-ml volumetric flask and dilute up to the mark with fresh AFW.

$$1 \text{ L } NH_3 - N \text{ in standard solution} \equiv 10 \text{ mg/L}$$

4. Standardization

Prepare a standard in duplicate with 500-ml standard NH_4Cl solution and follow the same procedure as described below for the sample.

5. Procedure

a. Preparation of distillation apparatus

i. Into four Kjeldahl distillation flasks measure 500 ml fresh AFW. Add 20 ml borate buffer to each flask.

ii. Adjust pH of solution in each flask to 9.5 with 6 N NaOH solution.

iii. Add a few clean, dry glass beads to the flasks to control bumping.

iv. Place all the flasks on heater and assemble the Kjeldahl unit as shown in Fig. 9.2.

v. Bring the solution to a boil.

vi. The steam thus produced will clean and make the distillation apparatus free of ammonia before use.

vii. Continue this process until the distillate is completely free of ammonia.

viii. Repeat this procedure each time if the apparatus is not in use for a period of 4 h or more, which is enough to accumulate ammonia in the apparatus.

b. Preparation of sample

i. If the sample contains residual chlorine, add a quantity of dechlorinating reagent in proportion to the residual chlorine present in the sample in accordance with following relation:

$$1\,\text{ml of dechlorinating reagent} \equiv 1\ \text{mg/L residual chlorine}$$

ii. The volume of sample for distillation should be selected according to Table 9.1.

TABLE 9.1
Volume of Sample Taken for Distillation in the Analysis of NH_3 – N

Range of NH_3 – N in sample (mg/L)	Volume of sample, ml
5 – 10	250
10 – 20	100
20 – 50	50
0 – 100	25

c. Distillation of sample

i. Measure 500 ml dechlorinated sample or an aliquot diluted to 500 ml with AFW in distillation flask.

ii. Adjust pH of sample to neutral with neutralizing solution, i.e., either 1 N NaOH or H_2SO_4 as required.

iii. Add 25 ml borate solution and adjust pH to 9.5 with 6 N NaOH.

iv. Measure 50 ml indicating-boric-acid reagent in receiver. Immediately start distillation.

Note: *Distillation of sample must start immediately after steam cleaning of distillation assembly to avoid contamination.*

v. The rate of distillation shall be adjusted in the range of 6–10 ml/min.

vi. Collect the distillate in the receiver.

Note: *The tip of delivery tube shall be submerged in the indicating boric acid solution placed in the receiver.*

vii. Collect about 300 ml distillate. Lower the receiver (no contact with tip of delivery tube). Continue distillation for 1 or 2 min more to clear the condenser and delivery tube. Replace the receiver with a conical flask.

viii. Continue distillation to concentrate the residue left in Kjeldahl flask (approximately to 50-ml volume). Collect the distillate in a conical flask.

ix. Do not preserve the excess distillate.

x. Preserve the residue left in the distillation flask for organic nitrogen determination, if required.

xi. Use distillate collected in the receiver for the estimation of $NH_3 - N$.

d. Estimation of $NH_3 - N$

i. Dilute 300 ml distillate to 500 ml with AFW.

ii. Titrate it with standard 0.02 N H_2SO_4 titrant until a pale lavender color is developed. This is the end point.

6. Calculation

$$\text{mg } NH_3 - N = \frac{(A - B) \times N \times E \times 1000}{\text{volume of sample, ml}}$$

where A = volume of H_2SO_4 titrant used for sample, ml
B = volume of H_2SO_4 titrant used for blank, ml
N = normality of H_2SO_4 titrant
E = equivalent mass of N

Since, $N = 0.02$ and $E = 14$, the above relation can be modified as follows:

$$\text{mg } NH_3 - N = \frac{(A - B) \times 0.02 \times 14 \times 1000}{\text{volume of sample, ml}} = \frac{(A - B) \times 280}{\text{volume of sample, ml}}$$

9.3 Nitrate – Colorimetric Method – Detectable Range: 0.1 to 2.0 mg/L $NO_3 - N$

For Source and Significance, see Section 9.1.

Measurement

1. Principle

The nitrate content present in the sample produces yellow color on reaction with brucine solution under acidic conditions. The absorbance of color is measured at 410 nm. The intensity of color is directly proportional to the amount of NO_3 present in the sample.

Note: *a. This estimation reaction is an exothermic reaction, i.e., heat is generated during the reaction. The estimation of NO_3 is significantly affected by heat generated during the reaction, so the reaction temperature must be controlled during the estimation. b. This procedure can be used for water samples with high salinity also.*

2. *Apparatus*

a. U V Spectrophotometer – Any commercially available model set at 410-nm wavelength.
b. Mechanical pipettes – with varying capacities
c. Hot water bath – to maintain a temperature of sample flask between 90 and 95°C
d. Cool water bath — maintain water temperature at 10–20°C
e. Volumetric flasks – 100 ml capacity
f. Filtration unit – as described in Section 6.6.1.
g. Test tubes – Pyrex glass

3. *Reagents*

a. Nitrate-free water (NFW): Use DDW to prepare all the reagents and solutions (Section 3.3.3).
b. Dechlorinating reagent, sodium arsenite solution, $NaAsO_2$: Weigh 0.5 g $NaAsO_2$ and transfer to a 100-ml volumetric flask. Dissolve in NFW. Dilute up to the mark.
c. Brucine — sulfanilic acid solution: Weigh 1 g brucine sulfate and 0.1 g sulfanilic acid and transfer both chemicals to a 100-ml volumetric flask. Dissolve in approximately 70 ml hot NFW. Now add 3.0 ml concentrated HCl to the flask. Cool the solution to room temperature and bring up the final volume to 100 ml.

Note: *(1) Brucine sulfate and sodium arsenite are toxic in nature, hence avoid ingestion or any contact with skin.*

(2) Sulfanilic acid is used to remove the interference caused by nitrites (NO_2) present in the sample up to 0.5 mg NO_2 – N/L level.

d. Sulfuric acid solution, 4:1 H_2SO_4: Measure 100 ml NFW in a 1-L beaker. Carefully add 400 ml concentrated H_2SO_4 to NFW. Cool the solution to room temperature.

Note: *This reaction is highly exothermic, so this dilution must be done in a fume hood.*

e. Sodium chloride, NaCl: Weigh 300 g NaCl and transfer to a 1L volumetric flask. Dissolve in NFW and dilute to 1 L volume.
f. Stock nitrate solution, anhydrous KNO_3:
   i. Dry an adequate quantity of anhydrous KNO_3 in an oven set at 103°C for 1 h. Cool to room temperature in a desiccating cabinet.

ii. Weigh accurately 7.218 g dry and cool KNO_3 and transfer to a 1-L volumetric flask. Dissolve in NFW. Finally, bring its volume to 1 L with NFW.

$$1 \text{ ml of stock solution} \equiv 1 \text{ mg } NO_3^- - N$$

Note: *This solution is stable for six months.*

g. Standard Nitrate solution: Measure 10 ml of stock nitrate solution in a 1-L volumetric flask. Dilute up to the mark with NFW.

$$1 \text{ ml of standard solution} \equiv 0.01 \text{ or } 10 \ \mu g \ NO_3^- - N$$

4. Standardization

Prepare calibration standards containing 0.5, 1.0, 1.5 and 2.0 mg/L NO_3 – N and standardize as follows:

a. Into four 100-ml volumetric flasks, measure 5, 10, 15 and 20 ml standard nitrate solution.
b. Dilute solution in each flask with NFW up to the mark.
c. Divide 10-ml volume of each standard solution into four different test tubes.
d. Use the same procedure for development of color as described below for the sample.
e. Prepare a blank with NFW instead of sample and set the spectrophotometer to zero at 410 nm.
f. Draw a calibration curve of absorbance against concentration of NO_3- in mg/L.

5. Sample Preparation

a. If sample contains suspended matter, filter it through GF/C filter paper.
b. If sample contains residual chlorine, add 0.05 to 0.10 ml (1 or 2 drops) of sodium arsenite solution to neutralize.

6. Procedure

a. Begin with clean, dry test tubes.
b. Measure 10-ml sample or an aliquot diluted to 10-ml in a test tube. Always prepare each sample in duplicate.
c. Prepare a blank using 10 ml NFW instead of sample.
d. Add 2 ml NaCl solution to each tube containing sample. Mix thoroughly by swirling.
e. Add 10-ml of 4:1 H_2SO_4 solution. Mix again thoroughly.
f. Cool test tubes in a water bath containing water at 10 to 20°C.

Note: If at this point any turbidity appears, take the sample in spectrophotometric cell and read its absorbance at 410 nm. Record this value, as it represents the sample blank.

g. Add 0.5-ml brucine-sulfanilic acid reagent. Swirl the tubes thoroughly for uniform mixing. Avoid the formation of air bubbles.
h. Place test tubes in hot water bath maintained at 90–95°C for 20 min.
i. After 20 min remove the tubes and place in the cool water bath maintained at 10–20°C .
j. When the thermal equilibrium is attained, remove the tubes from the water bath.
k. Set zero with DDW blank or sample blank.
l. Read the absorbance of yellow color developed in a spectrophotometer set at 410 nm against blank or sample blank.
m. Read the concentration of $NO_3 - N$ directly from the standard curve.

7. Calculation

$$\text{mg } NO_3^- - N/L = \frac{\mu g \; NO_3 - N}{\text{volume of sample, ml}}$$

$$\text{mg } NO_3/L = \text{mg/L nitrate N} \times 4.43$$

where

$$4.43 = \frac{\text{equivalent weight of } NO_3}{\text{equivalent weight of N}} = \frac{62}{14}$$

9.4 Nitrite – Diazotization Method — Detectable Range: 1.0 to 300 μg / L NO_2 – N

For Source and Significance, see Section 9.1.

Measurement

1. Principle

Nitrite present in water or wastewater undergoes diazotization on reacting with sulfanilamide under acidic conditions. The product is then coupled with N–(1–naphthyl) ethylenediamine dihydrochloride to produce a reddish-purple azo dye. The absorbance of azo dye is measured on a spectrophotometer set at 550 nm.

2. Precautions

a. Nitrite in the sample should be measured immediately after sample collection to avoid its microbial conversion to NO_3 or NH_3. Even if the sampling bottle is sterilized, the natural bacterial life in water may cause this conversion.

b. If samples are to be stored, preserve the sample as described in Section 3.2.1. The estimation should be done within no more than 24 h of sample collection.

3. Apparatus

a. U. V. spectrophotometer – Any commercially available model set at 550 nm.

b. Mechanical pipettes – with varying capacity

c. Volumetric flasks – 100 ml capacity

d. Beaker – 50 ml capacity

e. Filtration unit – as described in Section 6.6.1.

4. Reagents

a. Nitrite-free water (NFW): Use DDW for this purpose (see Section 3.3.3).

b. Hydrochloric acid, HCl, 20% (v/v) solution: Measure 50 ml of NFW in a 100-ml volumetric flask. Carefully add 20 ml concentrated HCl. Dilute up to the mark with NFW.

c. Sulfanilamide solution: Weigh 0.5 g sulfanilamide and transfer to a 100-ml volumetric flask. Dissolve in minimum amount of 20% (v/v) HCl. Make up the final volume to 100 ml.

d. Hydrochloric acid, HCl; 1% (v/v) solution: Add 1.0 ml concentrated HCl in a 100-ml volumetric flask containing 50 ml NFW. Dilute up to the mark with NFW.

e. N– (1– Naphthyl) ethylenediamine dihydrochloride solution: Weigh 0.3 g of the solid reagent and transfer to a 100-ml volumetric flask. Dissolve in 1% (v/v) HCl solution. Make up the final volume to 100 ml with 1% (v/v) HCl.

f. Aluminum chloride, $AlCl_3$: Refer to Section 7.5.8.

g. Stock nitrite solution:

   i. Place an adequate quantity of potassium nitrite, KNO_2, in a 50-ml beaker and keep for 24 h for drying in a desiccator containing concentrated H_2SO_4 (sp. Gr. 1.84).

   ii. Weigh accurately 0.6072 g dried KNO_2 and dissolve in minimal volume of NFW with stirring.

   iii. Transfer the solution to a 1L volumetric flask and make the volume up to the mark with NFW.

   iv. Preserve the solution with 2 ml chloroform.

   v. Store the solution in a sterilized Pyrex bottle under refrigeration.

$$1.0 \text{ ml stock nitrite solution} \equiv 100 \ \mu g \ NO_2 - N$$

Note: *(1) Potassium nitrite is readily oxidized in the moisture, so fresh bottles of reagent must be used for the preparation of stock solution.*

(2) If properly stored, the stock solution is stable for about 1 month.

(3) To achieve more accuracy, determine the nitrite content in stock solution before use by adding an excess of standard potassium permanganate, $KMnO_4$. The color of $KMnO_4$ is discharged with a standard reductant, ferrous ammonium sulfate, and, finally, back-titrating with standard permanganate solution.[4]

g. Standard nitrite solution: Measure 1.0 ml of stock nitrite solution in a 100-ml volumetric flask. Dilute to 100 ml with NFW.

$$1.0 \text{ ml standard nitrite solution} \equiv 1.0\ \mu g\ NO_2 - N$$

Note: *This solution is unstable. Always prepare fresh standard solution for each estimation.*

5. Standardization

Prepare calibration standards containing 1, 10, 30, 50, 100, 200 and 300 μg NO_2 – N/L concentration and calibrate by following procedure:

a. Into seven different 100-ml capacity volumetric flasks, measure 0.1, 1, 3, 5, 10, 20 and 30 ml standard nitrite solution.
b. Dilute solution in each flask up to the mark with NFW and mix thoroughly.
c. Prepare a blank in duplicate with NFW instead of standard solution.
d. Develop color in standards and blank as described below for the sample.
e. Record the absorbance at 550 nm and plot a calibration curve of absorbance against the concentration of NO_2 – N μg/L.

6. Procedure

a. Removal of turbidity and color:
   i. If the sample contains suspended solids, add a few drops of aluminum hydroxide, $Al(OH)_3$ suspension per 100-ml sample.
   ii. Stir thoroughly and allow to stand until the flocs settle down.
   iii. Filter through GF/C filter paper.
   iv. If coloration remains, prepare a sample blank. To 10-ml sample add 1-ml sulfanilamide and record the absorbance at 550 nm.
b. Development of color:
   i. Measure 100-ml sample or an aliquot diluted to 100 ml into a 250-ml Erlenmeyer flask. The temperature of sample must be maintained at room temperature (25°C).

ii. Add 2 ml of sulfanilamide solution and mix well. Wait 5 min for reaction.

iii. Add 2 ml of naphthyl ethylene diamine solution. At this stage, pH should be less than 2.0. Wait for 10 min.

iv. Measure the absorbance after 10 min in a spectrophotometer set at 550 nm against a DDW blank or sample blank.

v. Calculate the concentration of $NO_2 - N$ from the calibration curve.

6. Calculation

$$\mu g\ NO_2 - N/L = (A - B) \times df$$

where A = absorbance reading for sample
B = absorbance reading for blank or sample blank
df = dilution factor

If 10 ml of sample is diluted to 100 ml, df = 10.
If 20 ml of sample is diluted to 100 ml, df = 5.
If 1.0 ml of sample is diluted to 100 ml, df = 100.

$$mg/L\ NO_2 = mg/L\ NO_2 - N \times 3.29$$

where

$$3.29 = \frac{\text{equivalent weight of } NO_2}{\text{equivalent weight of N}} = \frac{46}{14}$$

9.5 Total Organic Nitrogen – Nesslerization Method

For Source and Significance, see Section 9.1.

Measurement

1. Principle

Nitrogen present in the organic constituents of a water or wastewater sample is converted to ammonium sulfate $[(NH_4)_2SO_4)]$ during digestion of the sample. Digestion is carried out with strong acid (H_2SO_4) in the presence of catalysts (potassium sulfate, K_2SO_4 and mercuric sulfate, $HgSO_4$). During digestion, the ammonium sulfate forms a mercury ammonium complex that is decomposed by sodium thiosulfate, $Na_2S_2O_3$ to ammonia. Ammonia is then distilled from an alkaline medium and absorbed in boric acid. Finally, the ammonia level is measured by Nesslerization to calculate the amount of organic N present in the sample.

2. *Apparatus*

a. Digestion unit – A semi-micro Kjeldahl digestion unit comprising Kjeldahl digestion flasks of 100-ml capacity is used. The heating device of this unit provides a temperature range of 365 to 380°C for effective digestion. The heater needs six places to accommodate six digestion flasks. This unit is commercially available.

Note: *This unit must be installed in a fume hood.*

b. Kjeldahl distillation apparatus – As described in the section 9.2.
c. Spectrophotometer – Any commercially available model adjusted at 425-nm wavelength.
d. Magnetic stirrer
e. Volumetric flasks of different capacities
f. Titration assembly – As described in Section 7.2.

3. *Reagents*

Prepare all the reagents in fresh ammonia-free water (AFW).

a. Ammonia free water (AFW): see section 9.2.
b. Sulfuric acid, 1:4 solution: Measure 80 ml fresh AFW in a 100-ml volumetric flask. Add carefully 20 ml concentrated H_2SO_4 and cool the solution.
c. Mercuric sulfate solution: Weigh 4 g red mercuric oxide (HgO) and transfer to a 50-ml volumetric flask. Dissolve in minimal volume of 1: 4 H_2SO_4 solution. Bring the final volume to 50 ml with fresh AFW.
d. Digestion mixture:
   i. Weigh 67 g potassium sulfate K_2SO_4 and transfer to a 500-ml volumetric flask. Dissolve in 300 ml fresh AFW. Then carefully add 100 ml concentrated H_2SO_4.
   ii. Keep the flask on a magnetic stirrer and add 20 ml mercuric sulfate ($HgSO_4$) solution with constant stirring.
   iii. Bring the final volume to 500 ml with fresh AFW.
   iv. To prevent crystallization, store the digestion solution at room temperature.
e. Alkaline sodium thiosulfate solution: Weigh 250 g NaOH and 12.5 g sodium thiosulfate ($Na_2S_2O_3$. 5 H_2O). Transfer both chemicals to a 500-ml volumetric flask. Dissolve in fresh AFW and dilute up to the mark.
f. Boric acid solution: Weigh 10-g boric acid and transfer to a 500-ml volumetric flask. Dissolve in fresh AFW and bring the final volume to 500 ml.
g. Nesslerís reagent:
   i. Solution A: Weigh 10 g mercuric iodide (HgI_2) and 7-g potassium iodide (KI) in a beaker and dissolve in a small volume of fresh NFW.
   ii. Solution B: In an another beaker, weigh 16 g NaOH and dissolve in 50 ml fresh NFW by keeping it on a magnetic stirrer. Cool the solution.
   iii. Add solution A to solution B and mix thoroughly.

iv. Transfer this mixture to a 100-ml volumetric flask and bring the final volume to 100-ml with fresh NFW.

Note: Store Nesslerís reagent in a Pyrex glass bottle away from direct sunlight. Stored carefully in a dark ,cool place it remains stable for more than six months.

h. Stock ammonium chloride, NH_4Cl solution:

i. Dry an adequate quantity of NH_4Cl in an oven set at 105°C for 1 h. Cool to room temperature in a desiccating cabinet.

ii. Weigh accurately 1.91 g dry, cool NH_4Cl and transfer to a 500-ml volumetric flask. Dissolve in fresh AFW. Dilute up to the mark with AFW.

$$1 \text{ ml stock } NH_4Cl \text{ solution} \equiv 1 \text{ mg } NH_3 - N$$

i. Standard NH_4Cl solution:

i. Standard I: Measure 10-ml stock NH_4Cl solution in a 100-ml volumetric flask. Dilute to 100 ml with fresh AFW.

$$\text{Concentration of } NH_3 - N \text{ in standard I solution } = 100 \text{ mg/L}$$

ii. Standard II: Measure 1.0 ml of standard I in a 100-ml volumetric flask and dilute with fresh AFW up to the mark.

$$\text{Concentration of } NH_3 - N \text{ in standard II solution } = 1 \text{ mg/L}$$

4. Standardization

a. Preparation of a calibration curve:

Prepare calibration solutions containing 1, 2, 4, 6, 8 and 10 mg NH_3 – N/L and standardize in the following manner:

i. Take six different 100-ml volumetric flasks and measure 1, 2, 4, 6, 8 and 10 ml of standard I respectively.
ii. Bring the volume in each flask to 100 ml with fresh AFW.
iii. Add 2 ml of Nesslerís reagent to each flask and mix thoroughly.
iv. Wait for 20 min.
v. Prepare a blank with 100 ml NFW instead of standard solution. Set zero with blank.
vi. Read the absorbance of standards against blank at 425 nm.
vii. Plot the values of absorbance against the concentration of NH_3 – N in mg/L to obtain a calibration curve.
viii. Periodically examine the calibration curve with fresh standard NH_4Cl solution.

b. Standard NH_4Cl solution:

   i. Take 50-ml of each standard I and standard II in two digestion flasks.

   ii. Follow the same procedure of digestion, distillation and estimation as described in Step 5.

5. Procedure

a. Place 50 ml of concentrated residue left during distillation procedure for the estimation of ammonia (see Section 9.2) into a micro-Kjeldahl digestion flask.

b. Add 10 ml of digestion mixture and few boiling chips or glass beads into the flask.

c. Place the flask on the Kjeldahl heating unit in a fume hood. Turn on the heater.

d. First evaporate the mixture on low temperature to reduce the volume to half.

e. Increase the temperature and evaporate the remaining volume until SO_3 fumes are given off. At this stage, the solution turns colorless or pale yellow.

f. Continue the process for 30 min more, turn off the heater and cool the residue.

g. Suspend the residue in 50 ml fresh NFW and transfer into an 800-ml Kjeldahl distillation flask.

h. Rinse the digestion flask with 30-ml of fresh AFW. Transfer the rinsing to a distillation flask. Make up the final volume to 300 ml with AFW.

i. While the process of digestion is being carried out, steam out the distillation unit as explained in Section 9.2.

j. Measure 50 ml of 2% boric acid into a marked receiver. Position the flask so that the tip of the condenser or delivery tube can extend nearly to the bottom of the receiver.

k. Carefully add 10 ml of alkaline sodium thiosulfate solution to the distillation flask to make the digestate alkaline.

Note: *(1) Slow addition of heavy alkaline solution is essential because it forms a layer below the aqueous H_2SO_4 solution. This prevents the loss of free $NH_3 - N$.*

(2) Do not swirl the flask to mix the solution until the distillation flask has been connected to the distillation apparatus.

l. Connect the Kjeldahl distillation flask to the condenser. Swirl to mix the solution in the flask.

m. Turn on the heater and distill the solution at a rate of 5 to 10 ml/minute. Continue the distillation until 300 ml of distillate will be collected in the receiver containing 50 ml boric acid solution.

n. Continue distillation until the volume of distillate reaches the 350-ml mark.

o. Remove the receiver and cover with aluminum foil.

p. Measure 50-ml sample of distillate or standard solution in a 250-ml Erlenmeyer flask. Add 2 ml of Nessler's reagent.

q. Wait 20 min after addition of Nessler's reagent.

r. Prepare a blank with 50 ml AFW and carry out all the steps of procedure as described for the sample

s. Read the absorbance against blank in a spectrophotometer set at 425 nm.

7. Calculation

$$\text{Total Organic N, mg/L} = \frac{A \times B \times 1000}{C \times \text{volume of sample, ml}}$$

where A = concentration of $NH_3 - N$ obtained from calibration curve
B = final distillate volume, ml
C = distillate volume taken for Nesserlization, ml

Note: *Total Kjeldahl Nitrogen (TKN)*

Total Nitrogen content in water or wastewater is presented as Total Kjeldahl Nitrogen (TKN). TKN measures free ammonia and NH_3 *produced by the conversion of nitrogen containing organic compounds like amino acids and proteins.*

To estimate TKN take unfiltered sample for digestion instead of taking residue left during distillation in the estimation of $NH_3 - N$.

9.6 Phosphorus

1. General Discussion

Phosphorus is an essential nutrient required in significant amounts in the synthesis and decay of organic matter. Phosphorus bounded organically has two major functions:

a. It is an essential structural element of nucleic acids.

b. It helps in the production, storage and utilization of chemical energy during vital activities.

Source

Phosphorus is a constituent of many inorganic and organic compounds widely distributed in the environment. It occurs mainly in the form of phosphates. The availability of phosphorus in natural waters and wastewater is largely due to the discharge of man-generated wastes and runoff. Major sources of phosphorus include fertilizers, boiler water conditioners, drinking water treatment aids, detergents and other laundering products. Organic phosphates are mainly produced in different biological processes. They are released to wastewater through animal wastes, food residues and effluents of animal-and-plant processing industries.

Significance

Phosphorus is the key nutrient along with N. It is required for the growth and activity of biological systems in the activated sludge process used for the treatment of wastewater. If the level of P in water and wastewater effluent is not monitored, it will give rise to the following adverse conditions:

1. It stimulates the growth of algae and other aquatic plants, which may cause eutrophication in water streams.
2. It interferes with the coagulation process in water treatment.
3. It interferes with lime – soda softening of water.

2. Separation of Different forms of Phosphorus – Digestion process

Phosphorus in water and wastewater generally occurs in the form of Ortho-phosphates, polyphosphates (meta-, pyro- or other polyphosphates) and organic phosphates. One of the most reactive forms is ortho-phosphate, which includes phosphate (PO_4^{3-}), hydrogen phosphate (HPO_4^{2-}), dihydrogen phosphate ($H_2PO_4^{-}$) and phosphoric acid (H_3PO_4). Polyphosphates include the molecules with 2 or more P atoms forming complex molecules. Organic phosphates are the products of various biological degradation processes.

Ortho-phosphate can be determined directly with ammonium molybdate. All other forms of phosphorus must be converted into ortho-phosphorus before estimation. The digestion of the sample under different conditions can do this separation.

The separation of phosphorus into various forms is carried out in following steps:

a. Suspended phosphates: The sample is filtered through a 0.45-μm membrane filter. This separates the dissolved forms of P from suspended or insoluble phosphates.
b. Ortho-phosphates: The sample is taken directly without filtration and digestion for analysis with ammonium molybdate. This result represents the content of P available as Ortho-phosphorus in the sample.
c. Hydrolyzable phosphorus with ortho-phosphorus: The sample is boiled with acid before estimation. The acid digestion converts hydrolyzable forms of P such as meta-, pyro- and tri-polyphosphates into ortho-phosphates. The result will present total ortho-phosphate content, including ortho-phosphate produced from hydrolyzable fraction as well as present in the undigested sample.
d. Total recoverable phosphorus: This is obtained by digesting the sample with acid-persulfate mixture. In this process, all forms of P including organic phosphates convert into Ortho-phosphates. Some heavy metallic phosphates remain undigested.

The complete analytical scheme to separate various forms of phosphorus is presented in Fig. 9.3.

9.7 Phosphorus – Ortho (Reactive) – Ascorbic Acid Method – Detectable Range: 0.1 to 2.5 mg/L PO_4^{3-}

For Source and Significance, see Section 9.6.

Measurement

1. Principle

Phosphorus present in water or wastewater in the form of ortho-phosphate is known as reactive phosphorus. It complexes with molybdate under acidic conditions producing a phosphomolybdate complex. Ascorbic acid then reduces this complex, producing an intense-blue-colored molybdenum complex. The absorbance of the color can be read at 890-nm wavelength.

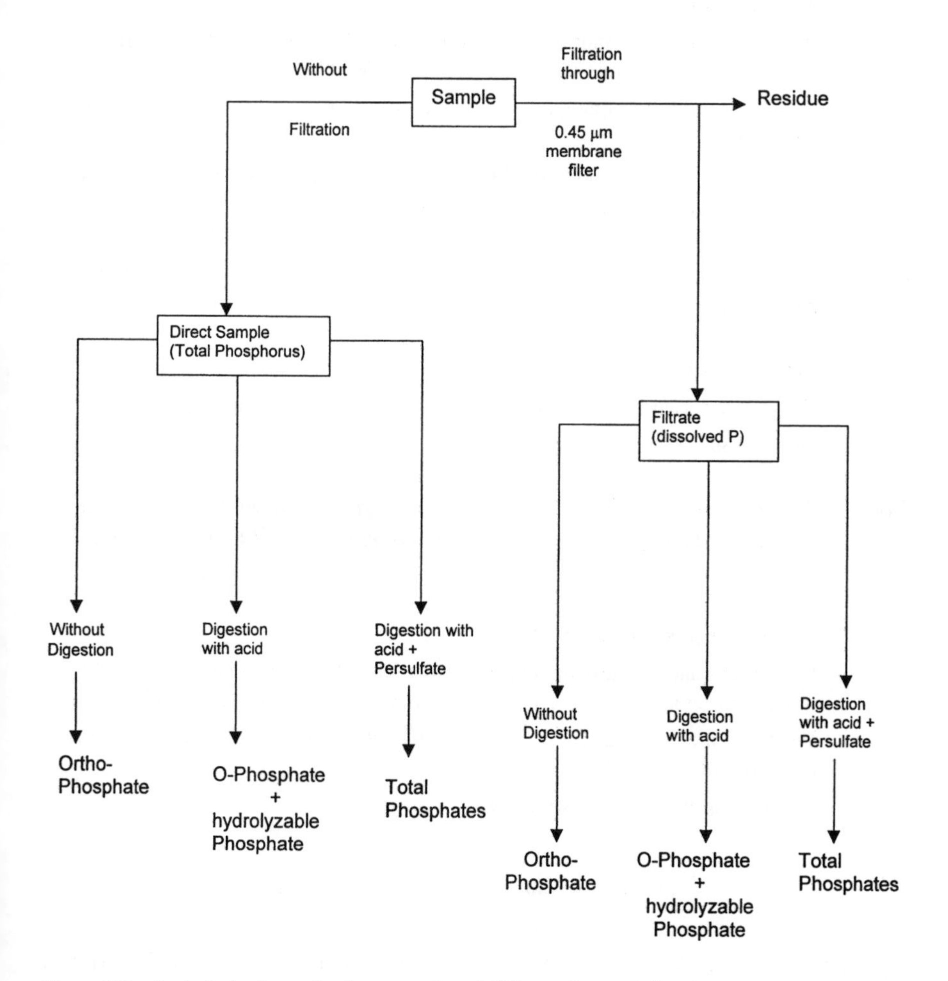

Figure 9.3 - Analytical scheme for the separation of different forms of phosphorus

2. *Apparatus*

a. U V Spectrophotometer – Any commercially available model set at 890 nm
b. Volumetric flasks – 50 ml and 1-L capacity
c. Erlenmeyer flasks – 100 ml capacity
d. Oven – set at 105°C
e. Desiccating cabinet

3. *Reagents*

a. Cleaning solution, HCl, 50% (v/v): Prepare 50% (v/v) HCl solution by adding 100 ml HCl to 100 ml DDW. Keep this solution for rinsing all glassware.
b. Sulfuric acid: concentrated
c. Complexing reagent:
   i. Weigh 0.13 g of antimony potassium tartarate [K (SbO) $C_4H_4O_6$. 0.5 H_2O] and transfer to a 1L volumetric flask. Dissolve in 500 ml DDW.
   ii. Add 5.6 g ammonium molybdate to the above solution and swirl until it dissolves completely.
   iii. Cautiously and slowly add 70 ml concentrated H_2SO_4 with continuous stirring.
   iv. Cool the solution and dilute it up to the mark with DDW.

Note: *Store the solution in a polyethylene bottle away from heat. This is stable for about 6 months.*

d. Combined reagent:
   i. Weigh 0.5 g ascorbic acid and transfer to a 100-ml volumetric flask.
   ii. Dissolve in 100 ml complexing agent see step c.

Note: *This reagent is stable only for a week under refrigeration. Be very particular about the date of preparation of this reagent. It should be clearly marked on the bottle.*

e. Phosphorus stock solution, KH_2PO_4:
   i. Dry an adequate quantity of anhydrous potassium dihydrogen phosphate (KH_2PO_4) in an oven set at 105°C for 1 h. Cool to room temperature in a desiccating cabinet.
   ii. Weigh accurately 0.4394 g dry and cool KH_2PO_4 and transfer to a 1L volumetric flask. Dissolve in DDW.
   iii. Dilute up to the 1L mark with DDW.

$$1.0 \text{ ml stock solution} \equiv 0.1 \text{ mg } PO_4^{3-} - P$$

f. Phosphorus standard solution: Measure 5 ml phosphorus stock solution in a 100-ml volumetric flask and bring the volume up to the mark with DDW.

$$1.0 \text{ ml standard solution} \equiv 0.005 \text{ mg } PO_4^{3-} - P$$

4. Standardization

Prepare calibration standards containing 0.5, 1.0, 1.5, 2.0 and 2.5-mg/L phosphate as follows and standardize:

a. Into five different 50-ml volumetric flasks and measure 5, 10, 15, 20 and 25 ml standard phosphorus solution.
b. Dilute the solution in each flask up the mark with DDW.
c. Transfer the standard solutions into five different Erlenmeyer flasks of 100-ml capacity.
d. Add 10 ml of combined reagent to each calibration solution. Swirl each beaker to mix.
e. Wait 20 min. for reaction. A blue color will develop.
f. In the meantime, prepare a blank with DDW instead of calibration solution.
g. After 20 min. read absorbance of standards at 890 nm against the Blank.
h. Plot a calibration curve of absorbance against concentration of Ortho-phosphate in mg/L at 890 nm.

5. Procedure

a. Rinse all the glassware with cleaning solution. Wash thoroughly with DDW.
b. Measure 50-ml sample into a 100-ml Erlenmeyer flask.
c. Add 10 ml of combined reagent and mixed thoroughly by swirling the flask several times.
d. Wait 20 min. for color development and then read the absorbance at 890 nm.

6. Important instructions

a. If the sample is turbid prepare a sample blank by adding the complexing reagent instead of combined reagent and record the absorbance.
b. If dissolved ortho-phosphate content is to be measured, filter the sample through 0.45-μm membrane filter and follow the procedure described above in step 4.

7. Calculation

Measure the ortho-phosphate content in the sample directly from the calibration curve.

$$\text{mg } PO_4^{3-} - P/L = \frac{(A - B) \times 1000}{\text{volume of sample, ml}}$$

where A = concentration of P in the sample from the curve, mg
B = concentration of P in the Blank from the curve, mg

9.8 Phosphorus – Total Recoverable – Digestion Method

For Source and Significance, see Section 9.6.

Measurement

1. Principle

The sample is digested with acid – persulfate mixture, which converts all the forms of P including organic and inorganic phosphates, to ortho-phosphates. After digestion, total P is measured quantitatively with the ascorbic acid procedure followed for ortho-phosphate estimation.

2. Apparatus

a. U V Spectrophotometer – set at 890 nm
b. Erlenmeyer flask – 125 ml capacity
c. Volumetric flask – 100 ml capacity
d. Graduate cylinder – 25 ml capacity
e. Hot plate
f. Spatula

3. Reagents

a. Cleaning solution, HCl, 50% (v/v): see Section 9.7.
b. Sulfuric acid, 9 N: see Section 3.5.3 and Table 3.4.
c. Sodium hydroxide, 2 N: see Section 3.5.3 and Table 3.5.
d. Phenolphthalein indicator: Use a commercially available ready-made solution.
e. Ammonium persulfate: Powder chemical.

Note: Ammonium persulfate is a strong oxidant, so handle with proper care.

4. Procedure

a. Digestion of the sample:
   i. Rinse all glassware with cleaning solution. After rinsing, wash thoroughly with DDW.
   ii. Measure 25-ml sample (without filtration) into a 125-ml Erlenmeyer flask.
   iii. Place some glass beads properly rinsed with cleaning solution and washed with DDW into the flask to prevent bumping.
   iv. Add approximately 0.5 g ammonium persulfate powder (can measure with a spatula scoop) into the flask. Swirl to mix.
   v. Add 1.0-ml of 9 N H_2SO_4 solution.

vi. Cover the flask with aluminum foil.

vii. Place the flask on a hot plate. Boil gently for 30 min.

Note: *Do not allow to boil dry. Maintain the volume nearly 20 ml by adding small volumes of DDW. The volume should not exceed 20 ml.*

viii. After 30-min. digestion, cool the sample to room temperature.

ix. Add 4–5 drops of phenolphthalein indicator and start adding 2 N NaOH solution dropwise till a pink color appears.

x. Add H_2SO_4 solution slowly until the pink color is discharged.

xi. Pour the sample into a 25-ml graduated cylinder and adjust the volume to 25 ml. Use this solution for analysis of phosphate.

b. Estimation

To estimate the total phosphorus content in the sample A follow the procedure described in Section 9.7.

5. Important Instructions

a. Digestion of the sample can be carried out in an autoclave for 30 min. at 121°C and 15 to 20 psi pressure.

b. If turbidity appears in the sample after digestion, prepare a sample blank and measure the absorbance. Deduct this value from the final reading of absorbance.

Chapter

Determination of Organic Constituents

Organic constituents in water streams may come from natural sources or by way of various human activities. Most of the natural organic compounds are soluble in water, but synthetic organics are insoluble in water. Organic constituents are classified into two categories:

1. Biodegradable Organics

Organic materials that are decomposed by microbes and utilized as food to get energy for their activity and growth are known as biodegradable organics. These materials usually consist of starches, fats, carbohydrates, proteins, alcohol, acids, aldehydes and esters.

The decomposition of biodegradable organic matter can be accomplished by following two processes:

a. Oxidation: Microbial decomposition of organics in presence of oxygen. This results in stable and acceptable products. This microbial process is known as Aerobic Decomposition.

b. Reduction: Removal of oxygen, present in its combined state in compounds either organic or inorganic in origin, to oxidize the organic matter. This process results in objectionable and malodorous products such as ammonia, hydrogen sulfide, mercaptans etc. This process is known as anaerobic decomposition.

Biological oxygen demand (BOD) is the representation of biodegradable organic content present in water or wastewater sample.

2. Non Biodegradable Organics

The organic materials resistant to biological decomposition are called non-biodegradable organics. The following constituents fall under this category:

a. Constituents of woody plants: Tannin, lignin, cellulose and phenols are the constituents of woody plants, which biodegrade so slowly that they are usually considered non-biodegradable organics.

b. Benzene-containing compounds: Detergents and petroleum products containing benzene and its derivatives are essentially non-biodegradable components.

c. Toxic compounds: Pesticides, insecticides, industrial chemicals and hydrocarbons.

The measurement of chemical oxygen demand (COD) includes both biodegradable and non-biodegradable organic constituents of water or wastewater samples. BOD should be subtracted from COD to quantify the non-biodegradable organics.

10.1 Biochemical Oxygen Demand (BOD) - Dissolved Oxygen Dilution Method

Background

BOD is the representation of the amount of oxygen required for microbial decomposition, i.e., complete aerobic decay and mineralization of biodegradable organic matter present in water or wastewater. The term aerobic decomposition means the breakdown of complex organic molecules into their simpler constituents that can serve as food for the growth and activity of microbes and liberate energy in presence of dissolved oxygen (DO).

The aerobic decomposition of organics takes place in two stages:

a. Carbonaceous Oxidation: This is the first stage, in which mainly the carbonaceous (carbon fraction) matter is oxidized. Generally it reaches to maximum stabilization in five days.

$$\text{DO} + \text{organic matter} \xrightarrow[\text{Protozoa}]{\text{Bacteria and}} CO_2 + H_2O + \text{microbial cells} \tag{10.1}$$

b. Nitrogenous (non-carbonaceous) Oxidation: In this second stage, nitrogenous (nitrogen-containing) substances such as ammonia are attacked by bacteria and develop a demand of oxygen as shown in reaction 10.2.

$$\text{DO} + NH_3 - N \xrightarrow[\text{bacteria}]{\text{nitrifying}} NO_3 - N + \text{microbial cells} \tag{10.2}$$

For the purpose of water and wastewater treatment, the first-stage oxygen demand is considered BOD because the maximum carbonaceous components of organic matter are stabilized. The aerobic decay of carbonaceous matter occurs in multiple steps. At each, a part of organic matter is oxidized to CO_2 and H_2O and rest is utilized for the formation of new microbial cells. This aerobic decay during 5 days of incubation is presented in Fig. 10.1 assuming that microbial cell yield is

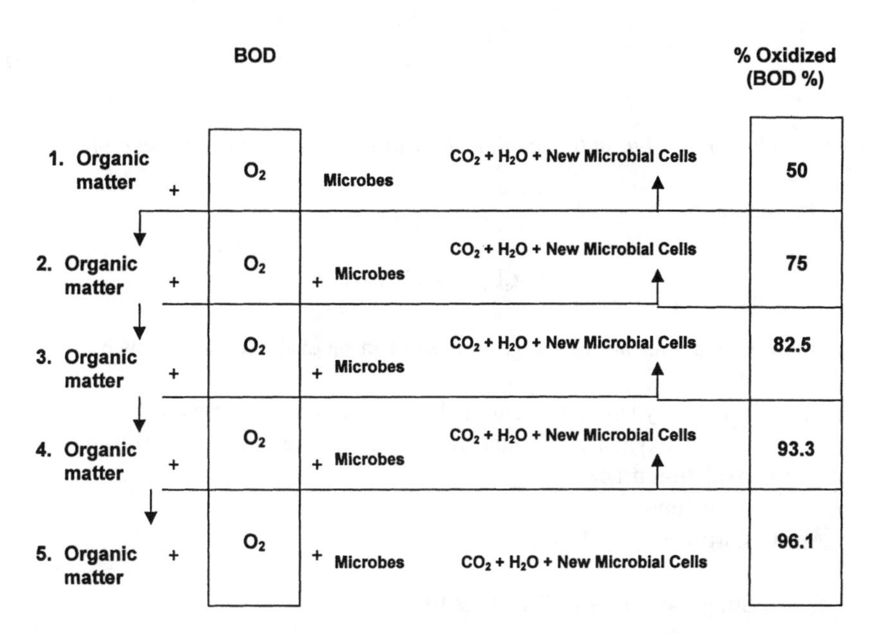

Figure 10.1 - Biological decay of organic matter under aerobic conditions. (Adapted from Gaudy, A. and Gaudy, E., *Microbiology for Environmental Scientists and Engineers*, McGraw-Hill, Inc., 1980, with permission).

50% and new microbial cells synthesized in each step can be metabolized by other microbial cells in the following step. Thus, in five days maximum (96%) carbonaceous matter is stabilized[30]. Hence, the term BOD is generally represented as BOD_5, which means the oxidation of organic waste in five days at 20°C.

The extent of oxidation of nitrogenous compounds during the five-day incubation depends on the presence of nitrifying bacteria. Usually during this period the nitrogenous oxidation does not begin because nitrifying bacteria are not available in sufficient number in raw sewage or primary effluent.

Kinetics of BOD

Kinetically, the rate of biochemical oxidation of organic matter is a first-order reaction, which means the rate of oxidation is directly proportional to the amount of unoxidized organic matter, i.e., the amount of food available to the microorganisms at any time.

$$\frac{-dL_t}{dt} \quad \alpha\, L_t$$

$$\frac{-d\,L_t}{d\,t} = k'\,L_t \tag{10.3}$$

$$\frac{d\,L_t}{L_t} = -k'\,d\,t \qquad (10.4)$$

Note: The minus sign indicates that the value of L_t decreases as time proceeds.

On integration:

$$\log_e L_t = -k'.t + C \qquad (10.5)$$

where C is a constant of integration and can be evaluated as follows:

If, L_t = oxygen equivalent of the organic present at time t
L_o = oxygen equivalent of the organic present at time 0, i. e., at the start of the oxidation process
t = time
At the start: t = 0 and $L_t = L_0$

Substituting the values in Equation 10.5:

$$\log_e L_0 = -k'(0) + C$$
$$\therefore C = \log_e L_0$$

Substituting the value of C in Equation 10.5 we get:

$$\log_e L_t = -k'.t + \log_e L_0$$
$$\log_e L_t - \log_e L_0 = -k'.t$$
$$\log_e \frac{L_t}{L_0} = -k'.t$$
$$2.3\log_{10}\frac{L_t}{L_0} = -k'.t$$
$$\log_{10}\frac{L_t}{L_0} = -k'.t/23 = 0.434\,k'.t$$

Let 0.434 k' = K where K is deoxygenation constant at given temperature, so

$$\log_{10}\frac{L_t}{L_0} = -K.t \qquad (10.6)$$

$$\therefore L_t = L_0 \cdot 10^{-k.t} \tag{10.7}$$

The oxygen equivalent remaining is not the parameter of concern, while the amount of oxygen consumed in oxidation, i.e., BOD, is the parameter of primary importance. If BOD is represented as Y_t the following relation exists :

$$Y_t = L_0 - L_t$$

$Y_t = L_0 - L_0 10^{-k.t}$ (substituting the value of L_t from Equation 10.7)

$$Y_t = L_0(1 - 10^{-k.t}) \tag{10.8}$$

Thus, Equation 10.8 represents the BOD exerted by the carbonaceous component of the organic compounds. The value of k for any given organic compound is temperature dependent as presented in Equation 10.9:

$$k_T = k_{20}[1.047]^{T-20} \tag{10.9}$$

where k_T = constant at temperature T°C
k_{20} = constant at temperature 20°C

For the determination of BOD at any temperature the value of k should be corrected for that particular temperature according to the relation given in Equation 10.9.

Significance

1. Assessment of pollution

The measurement of BOD is a direct representation of the extent of pollution in natural water bodies and potable water.

2. Wastewater treatment process

It is the most important parameter required for the design and operation of a wastewater treatment plant working on the principle of the activated sludge treatment process. It will represent the amount of food available for microbes to carry out oxidation of organic wastes.

3. Efficiency of treatment process

By comparing the BOD of influent and effluent of a wastewater treatment plant, the efficiency and effectiveness of the treatment process can be ascertained.

4. Quality of waste discharges

BOD also determines the quality of waste discharged to the streams because bacteria in the stream will oxidize the discharged organic matter and consume oxygen dissolved in water. Thus the oxygen dissolved in water depletes and affects the aquatic life available in the stream.

Measurement

1. Principle

Two methods are widely accepted for BOD measurement — dilution method and manometric method.

In this section the dilution method is described in detail. Samples of water and wastewater are prepared by placing various incremental portions of the samples in BOD bottles with and without dilution water. The dilution water contains a known amount of dissolved oxygen with inorganic nutrients and a buffer. The buffer will help to maintain the pH of the solution within optimum range for activity of microorganisms.

The bottles are completely filled and sealed without entrapment of any air bubbles. After measuring the initial DO content, the bottles are incubated for five days at constant temperature, 20 ± 1°C in a BOD incubator. During the incubation period, bacteria carry out the aerobic decomposition of organic matter using the DO available in the sample. At the end of five days, the remaining DO is measured.

The relationship between oxygen consumed and the volume of the sample taken is used to calculate BOD. DO level can be measured by the Winkler method or directly by DO meters available in the market. In this section, the Winkler method is described.

2. Apparatus

a. Aspirator bottle – 10 L capacity
b. Air Pump – To aerate the dilution water
c. BOD Incubator – Adjusted at 20 ± 1°C. Record the temperature daily.
d. BOD Bottles – 300-ml capacity
e. pH meter

3. Reagents

a. Dilution water

i. Phosphate buffer solution:

KH_2PO_4 (potassium dihydrogen phosphate)	8.5 g
K_2HPO_4 (dipotassium hydrogen phosphate)	21.75 g
$Na_2HPO_4 \cdot 12\ H_2O$ (disodium hydrogen phosphate)	44.6 g

or

$Na_2HPO_4 \cdot 7\ H_2O$	33.4 g
NH_4Cl (ammonium chloride)	1.7 g

Weigh all the ingredients and transfer to a 1L volumetric flask. Dissolve in DDW. Dilute to the mark. Final pH should be 7.4 ± 0.1.

Note: Discard the buffer if the pH changes from this range.

ii. Magnesium sulfate solution, $MgSO_4 \cdot 7\ H_2O$: Weigh 2.25 g $MgSO_4 \cdot 7\ H_2O$ and transfer to a 100-ml volumetric flask. Dissolve in DDW and dilute up to the mark with DDW.

iii. Calcium chloride solution, $CaCl_2$: Weigh 2.75 g $CaCl_2$, anhydrous or 36.4 g $CaCl_2 \cdot 2\ H_2O$ and transfer to a 100-ml volumetric flask. Dissolve in DDW and dilute up to the mark with DDW.

iv. Ferric chloride solution, $FeCl_3 \cdot 6\ H_2O$: Weigh 0.025 g $FeCl_3 \cdot 6\ H_2O$ and transfer to a 100-ml volumetric flask. Dissolve in DDW and dilute up to the mark with DDW.

b. Dechlorinating Agent:

i. Sodium sulfite solution, Na_2SO_3 anhydrous, 0.025 N: Weigh 0.788 g anhydrous Na_2SO_3 and transfer to a 500-ml volumetric flask. Dissolve in DDW. Dilute up to the mark. Prepare fresh solution for every estimation.

ii. Sulfuric acid solution, H_2SO_4, 2% (v/v): Measure 200 ml DDW in a 250-ml volumetric flask. Add carefully 5 ml concentrated H_2SO_4 to DDW. Dilute to the mark. Cool the solution.

iii. Potassium iodide solution, KI: Weigh 10 g KI and transfer to a 100-ml volumetric flask. Dissolve in DDW. Dilute up to the mark with DDW. This solution is stable for a month. Store in a dark place.

Note: Discard the solution if it is yellow.

iv. Starch indicator — refer to Section 7.5.8.4g

c. Neutralizing agents:

i. Sulfuric acid, H_2SO_4, 1 N: Refer to Section 3.5.3 and Table 3.4.

ii. Sodium hydroxide, NaOH, 1 N: Refer to Section 3.5.3 and Table 3.5.

d. Reagents for DO measurement:

i. Manganese sulfate solution, $MnSO_4 \cdot H_2O$: Weigh 36.4 g $MnSO_4 \cdot H_2O$ and transfer to a 100-ml volumetric flask. Dissolve in DDW and dilute up to the mark.

Note: Filter the reagent if any sediment settles at the bottom of the flask.

ii. Alkaline iodide - azide reagent:

Sodium hydroxide, NaOH	50.0 g
Sodium iodide, NaI	13.5 g
Sodium azide, NaN_3	1.0 g

Weigh the above-mentioned quantities of NaOH and NaI and transfer to a 100-ml volumetric flask. Dissolve NaN_3 in 10 ml DDW separately and add to alkaline KI solution. Make up the final solution to 100 ml.

iii. Sulfuric acid, H_2SO_4: Concentrated

$$1.0 \text{ ml conc. } H_2SO_4 \equiv 3.0\text{-ml alkaline iodide-azide solution}$$

iv. Sodium thiosulfate stock solution, $Na_2S_2O_3$. 5 H_2O, 0.10 N: Weigh 24.82 g $Na_2S_2O_3$. 5 H_2O and transfer to a 1L volumetric flask. Dissolve in DDW. Dilute up to mark. Preserve the solution by adding 5 ml chloroform.

v. Standard sodium thiosulfate titrant, 0.025 N: Measure 250-ml sodium thiosulfate stock solution to a 1L volumetric flask. Dilute up to the mark with DDW.

vi. Standard potassium dichromate, $K_2Cr_2O_7$, 0.025 N: see Section 7.5.8.4e

vii. H_2SO_4, 10% (v/v) – Measure 300 ml DDW in a 500-ml volumetric flask. Carefully add 50 ml concentrated H_2SO_4. Bring the volume up to the mark. Cool the solution.

viii. Potassium iodide, KI - Use KI crystals.

e. Standardization of $Na_2S_2O_3$ titrant: Follow procedure described in Section 7.5.8.5a

f. Seed

Seed is an active microbial population capable of oxidizing biodegradable organic matter. The most common source of seed material is domestic raw sewage, which is stored at 20°C for 24 to 36 h, and allowed to stand undisturbed until most solids settle before being used in the estimation. Pipette from the upper portion of the seed material. The addition of 3.0 ml of raw domestic sewage seed to each liter of dilution water is ample. Seed that has a BOD of 200 mg/L (normal value in domestic sewage) when added at the rate of 3 ml/L dilution water depletes 0.6 ppm DO.

4. Procedure

a. Preparation of dilution water

Take 5L DDW in aspirator bottle and add 10 ml each of reagent phosphate buffer, $MgSO_4$, $CaCl_2$ and $FeCl_3$. Fill aspirator bottle up to 10 L with DDW. Aerate the dilution water continuously with the help of an aeration pump. The dilution water must be saturated with O_2 and maintained at 20 ± 1°C.

Note: If dilution water is stored, add phosphate buffer just prior to use.

b. Test of dilution water quality

Fill a 300-ml BOD bottle with dilution water. Stopper tightly without entrapment of air bubbles. Incubate for five days at 20°C. Determine DO before and after incubation. The depletion of O_2 should not be more than 0.1 mg/L. If consumption exceeds this limit, prepare fresh dilution water for analysis.

c. Seeded dilution water

Use of seeded dilution water is not required for the analysis of wastewater, effluent of wastewater treatment plant (unless it has been chlorinated), or river water. However, some samples, such as some untreated industrial wastes, disinfected wastes, high temperature wastes, and wastes with extreme pH values do not contain enough bacterial population to oxidize any organic matter. Such types of wastes must be diluted with seeded dilution water to add some active bacterial population to the wastes. Normally, 2 ml of seed per l of diluted sample is used.

Note: *This applies to treated effluents in which original bacterial life has been destroyed by chlorination.*

d. DO measurement - modified Winkler method

i. Rinse a 300-ml BOD bottle with sample. Pour the sample into the BOD bottle in such a way that it is allowed to overflow to avoid any entrapment of air bubbles. Close the bottle with stopper carefully. The temperature of the sample should be recorded before filling the bottle.

ii. Remove the stopper and add 2 ml $MnSO_4$ reagent followed by 2-ml alkaline iodide – azide reagent. Replace the stopper so that no air is trapped in the bottle. Pour any excess water off the bottle rim and invert it several times to mix. A brownish-orange flocculent precipitate will form, if sample contains DO. In absence of DO, a white precipitate will appear.

iii. Allow the sample to stand until the floc has settled and left the top half of the solution clear. Again invert the bottle several times and let stand until the upper half of the solution is clear. This is to ensure the complete reaction of chemicals with available DO.

iv. Remove the stopper and immediately add 2 ml concentrated H_2SO_4. Replace the stopper carefully to avoid traping any air bubbles. Invert several times to mix. The floc will dissolve and leave a yellowish-orange iodine color if DO is present.

v. Measure 204 ml of sample, which corresponds to 200 ml of the original sample (correction for the sample loss by displacement with reagent) into a 250-ml Erlenmeyer flask.

Note: *For 300-ml sample, 2 ml of each reagent ($MnSO_4$, azide and H_2SO_4) is added. Hence, for 200-ml sample equivalent volume will be = 200 x 306 / 300 = 204 ml*

vi. Titrate this solution with standard 0.025 N $Na_2S_2O_3$ titrant to a faint yellow color. Add 1–2 ml of starch indicator and swirl to mix. A dark blue color will develop.

vii. Continue the titration until the solution changes from blue to colorless.

$$1.0 \text{ ml of } 0.025 \text{ N } Na_2S_2O_3 \equiv 200\ \mu g \text{ DO}$$

So the total number of ml $Na_2S_2O_3$ solution used is equal to the mg/L DO available in the sample.

e. Sample Pretreatment

i. Acidic or alkaline samples: They must be neutralized with 1N NaOH or 1N H_2SO_4 so that the reagent does not dilute the sample by more than 0.5%.

ii. Chlorinated samples: The residual chlorine available in chlorinated samples must be neutralized with 0.025 N Na_2SO_3. To determine the appropriate quantity of Na_2SO_3 required to neutralize residual chlorine, the following steps must be followed:

- Add 10 ml of 2% (v/v) H_2SO_4 solution followed by 10 ml of the KI solution to 100-ml sample.
- Add a few drops of starch indicator and swirl to mix. A blue color will appear.
- Titrate with 0.025 N Na_2SO_3 titrant till the solution becomes colorless.
- Calculate the amount of Na_2SO_3 titrant needed to dechlorinate the sample according to the following relation:

$$\text{Vol. of } 0.025 \text{ N } Na_2SO_3 = \frac{\text{ml used} \times \text{ml to dechlorinate}}{100}$$

- Add that amount of 0.025 N Na_2SO_3 standard solution that has been obtained after calculation to the fresh sample to neutralize residual chlorine. Mix thoroughly. Allow the sample to stand for 10–20 min. and proceed for BOD measurement.

iii. Supersaturated samples: Cold samples containing more than 9 mg/L DO are known as supersaturated samples. They must be brought to saturation state by bringing their temperature to about 20°C. This can be done by vigorously agitating the bottle partly filled with the sample for 2 min. or aerating with filtered compressed air for 2 h.

f. Sample dilution

i. Undiluted sample: Those samples that contain initial DO near saturation level and five day BOD less than 7 mg DO/L need not to be diluted.

ii. Diluted sample: The samples with suspected high BOD having no or very little DO level need to be diluted with dilution water to obtain the sufficient level of DO in the sample. Hence, a clear depletion of DO during five days, incubation can be achieved to obtain an appropriate value of BOD in the samples. In the absence of prior knowledge, the following data may help to decide about the need for dilution of the sample:

Strong industrial waste	0.1–1.0 %
Raw wastewater & primary effluent	1.0–5.0 %
Secondary effluent	5.0–25 %
Highly polluted surface water	25–100 %

g. Measurement of BOD

i. Undiluted sample

1. Fill two BOD bottles with dilution water and label them B_I and B_5 respectively.
2. For each sample prepare two bottles filled with sample. Label one S_I, which is used to measure initial DO in the sample. Label another S_5; this is to be kept for five days' incubation at 20°C.

3. Measure initial DO in BOD bottles labeled B_I and S_I according to the Winkler method. Record these values as $DO_{I\,(b)}$ and $DO_{I\,(s)}$ i.e., DO in blank and sample respectively.
4. Incubate BOD bottles labeled B_5 and S_5 in a BOD incubator for five days at 20 ± 1°C.
5. After five days' incubation, measure the DO in all bottles and record as $DO_{5(b)}$ and $DO_{5(s)}$.

ii. Diluted sample

1. Measure selected volume of sample with the help of wide-tip volumetric pipettes in BOD bottles.
2. Fill bottle with enough dilution water.
3. Seed the sample if necessary before filling with dilution water.
4. Carefully stopper the bottle without entrapment of any air bubbles.
5. If dilution required is more than 1:100, make a primary dilution in a graduated cylinder before making final dilution in the BOD bottle.
6. Prepare sets of blank and each sample as described under undiluted sample. Measure initial DO using Winkler method.
7. Incubate all bottles for five days at 20 ± 1°C.
8. Measure DO in all bottles and calculate BOD as described below.

iii. Seeded dilution water

Prepare the sample of seeded dilution water as described above. Record the initial DO as SD_I. Incubate the bottles at 20 ± 1°C for five days. Measure the DO and record it as SD_5.

5. Calculation

a Undiluted sample

$$\text{mg BOD/L} = DO_I - DO_5$$

where DO_I = initial Dissolved oxygen
DO_5 = final Dissolved oxygen after 5 days

b. Diluted sample: If dilution water is not seeded

$$\text{mg BOD/L} = \frac{DO_I - DO_5}{\rho}$$

where ρ = decimal fraction of sample used

c. Diluted sample: If dilution water is seeded

$$\text{mg BOD/L} = \frac{DO_I - DO_5 - (SD_I - SD_5)f}{\rho}$$

where SD_I = initial DO of seeded dilution water
SD_5 = DO of seeded dilution water after 5 days incubation

$$f = \frac{\%\text{ seed in diluted sample (approx. 0.2\%)}}{\%\text{ seed in dilution water blank}}$$

10.2 Chemical Oxygen Demand (COD) – Dichromate Reflux Method

The COD is considered mainly the representation of pollution level of domestic and industrial wastewater, or contamination level of surface, ground, and potable water. This is determined in terms of total oxygen required to oxidize the organic matter to CO_2 and water.

The COD values include the oxygen demand created by biodegradable as well as non-biodegradable substances, because it involves oxidation of organic matter with strong oxidizing chemicals. As a result, COD values are greater than BOD and may be much greater when significant amounts of biologically resistant organic matter is present.

Source

May come from natural sources or through various human activities.

Significance

Commonly COD is used to define the strength of wastewater containing non-biodegradable organic substances or compounds that inhibit biological activity. With COD estimation no differentiation can be made between biologically oxidizable and biologically inert organic matter. Hence, it provides no information related to the rate of biological stabilization of organic matter. This is the only limitation with COD estimation.

In comparison with BOD, COD estimation has an advantage in that it requires a short digestion period of about 3 h rather than five days' incubation period as required for BOD estimation. Hence, a rapid and frequent monitoring of treatment-plant efficiency can be maintained with COD measurement. If the nature of wastes received at treatment plant is almost constant, a relationship between BOD and COD can be established that helps the operator to monitor the activated sludge treatment of wastewater.

Measurement

1. Principle

The COD of wastewater or polluted water is a measure of the oxygen equivalent of the organic matter susceptible to oxidation by strong oxidizing chemicals, i.e., by a mixture of chromic acid and sulfuric acid. The organic matter is converted to CO_2 and H_2O by heating with potassium dichromate ($K_2Cr_2O_7$) in an acidic medium containing silver sulfate. The nascent O produced in the reaction Equation 10.10 oxidizes the organic matter as shown in Equation 10.11.

$$K_2Cr_2O_7 + 4\ H_2SO_4 \rightarrow K_2SO_4 + Cr_2(SO_4)_3 + 4\ H_2O + 3\ O \qquad (10.10)$$

$$\text{Organic matter} + O \rightarrow CO_2 + H_2O \qquad (10.11)$$

The dichromate reduces to Cr^{3+} state, which imparts green color to the reacting solution. The overall reaction is presented in Equation 10.12.

$$\text{Organic matter} + Cr_2O_7^{2-} + H^+ \rightarrow CO_2 + H_2O + Cr^{3+}\ \text{(green)} \qquad (10.12)$$

The excess of dichromate is titrated with ferrous ammonium sulfate (FAS) using ferroin indicator. The reaction between FAS and dichromate can be represented as Equation 10.13. First the green color appears due to reduction of $K_2Cr_2O_7$ to Cr^{3+} state. At end point, the green color discharges and the reddish-brown color of the indicator remains.

$$6\ Fe^{2+} + Cr_2O_7^{2-} + 14\ H^+ \rightarrow 6\ Fe^{3+} + 7\ H_2O + 2\ Cr^{3+}\ \text{(green)} \qquad (10.13)$$

Thus, the amount of oxidizable organic matter is proportional to the $K_2Cr_2O_7$ consumed.

2. Apparatus

a. Refluxing unit – comprises the following
   i. Erlenmeyer flask – 250 ml capacity with standard joints
   ii. Condenser – double jacketed with standard joints
   iii. Hot plate
b. Magnetic stirrer
c. Titration assembly – see Section 7.2.
d. Homogenizer / blender
e. Oven – set at 103°C
f. Desiccating cabinet
g. Dispensers – to deliver accurate volume of chemicals

3. Reagents

a. Cleaning solutions:
   i. Chromic acid: Refer to Section 3.4.1.
   ii. Sulfuric acid Solution, H_2SO_4, 50% (v/v): Refer to Section 3.5.3 and Table 3.4.

b. Mercuric Sulfate, $HgSO_4$: Use analytical grades $HgSO_4$.

c. Ferroin Indicator: Weigh 1.485 g 1,10-phenanthroline monohydrate and 0.695 g $FeSO_4$. 7 H_2O. Transfer both the chemicals to a 100-ml volumetric flask. Dissolve in DDW. Dilute up to the mark with DDW.

d. Potassium Dichromate Solution, $K_2Cr_2O_7$, 0.25 N: Dry an aliquot quantity of analar $K_2Cr_2O_7$ in an oven at 103°C for 2 h. Cool to room temperature. Accurately weigh 12.259 g dry, cool $K_2Cr_2O_7$ and transfer to a 1L volumetric flask. Dissolve in DDW. Add 0.12 g of sulfamic acid to the concentrated dichromate solution. Dilute up to the mark.

Note: *(1) Sulfamic acid is added to remove the interference caused by* $NO_2 - N$.

(2) Attach a dispenser to the reagent bottle so it can dispense 20 ml $K_2Cr_2O_7$ *accurately.*

e. Sulfuric acid, H_2SO_4: concentrated

f. Sulfuric acid concentrated with silver sulfate, H_2SO_4 - Ag_2SO_4 catalyst: Weigh 22 g of silver sulfate (Ag_2SO_4) and add to a 2.5L concentrated H_2SO_4 bottle. Keep this solution on a magnetic stirrer. Stir for 1–2 days for complete dissolution of $AgSO_4$.

Note: *Attach a dispenser to this bottle, which can deliver 30.0 ml of this solution accurately.*

g. Ferrous ammonium sulfate (FAS) solution, Fe $(NH_4)_2$ $(SO_4)_2$. 6 H_2O approx. 0.25 N: Weigh 98 g FAS and transfer to a 1L volumetric flask. Dissolve in about 500 ml DDW. Add 20-ml conc. H_2SO_4. Dilute to 1 L with DDW and cool it.

h. FAS titrant, 0.10 N: Measure 400 ml of the 0.25 N FAS solution in a 1L volumetric flask. Dilute to 1L with DDW. Standardize this solution daily before estimation.

4. Important Instructions for Cleaning

a. Cleaning of glassware is essential, since a minute fraction of organic contaminant will produce false results. Hence flasks, condensers, i.e., the complete refluxing unit, must be washed with 50% H_2SO_4, taking full precaution prior to each use.

b. Cleaning with chromic acid: Daily clean Erlenmeyer refluxing flasks with chromic acid to prevent deposit buildup on the walls.

The best method is to fill the refluxing flasks with chromic acid and leave overnight. Next day, collect the solution in the storing bottle for reuse. Wash the flasks thoroughly under running water and finally rinse with DDW.

c. Glass beads: Glass beads must be used during refluxing to prevent the bumping of solution. They must be soaked in 50% (v/v) H_2SO_4 cleaning solution. They must be thoroughly washed, rinsed with DDW and dried in an oven at 103°C before each use. Properly cleaned and dried glass beads must be stored in a covered bottle.

5. *Interference*

a. Certain reduced inorganic ions can be oxidized under conditions of COD estimation and cause interference in the analytical test. Chlorides especially interfere with the reaction, as follows:

$$6\ Cl^- + Cr_2O_7^{2-} + 14\ H^+ \rightarrow 3\ Cl_2 + 2\ Cr^{3+} + 7\ H_2O \qquad (10.14)$$

Mercuric sulfate, $HgSO_4$ is used to eliminate the interference caused by chlorides, which forms a poorly ionized mercuric chloride complex.

$$Hg^{2+} + 2\ Cl^- \leftrightarrow HgCl_2\ \ (k = 1.7 \times 10^{13}) \qquad (10.15)$$

In the presence of excess mercuric ions the Cl^- ion concentration is reduced so much that it is not oxidized by dichromate.

b. Nitrites also interfere with the reaction. This interference can be overcome by addition of sulfamic acid. However, a significant amount of nitrites is not available in wastewater, because of insufficient dissolved oxygen.

c. Silver sulfate is used as a catalyst to oxidize straight-chain organic compounds.

6. *Blanks*

COD analysis is designed to measure oxygen requirements for oxidation of organic matter present in the samples. Thus, no organic matter from external sources must be present in order to obtain the accurate oxygen requirement. This is almost impossible, hence a blank with DDW is prepared and refluxed along with the sample in every estimation.

7. *Standardization*

a. FAS titrant, 0.10 N
   - i. Fill the burette with 0.10 N FAS titrant.
   - ii. Accurately measure 10 ml of 0.25 N $K_2Cr_2O_7$ solution into a clean Erlenmeyer flask and add 90 ml DDW into the flask.
   - iii. Dispense 30-ml conc. H_2SO_4 with constant stirring and cool the solution.
   - iv. Add 0.5-ml ferroin indicator.
   - v. Titrate with FAS titrant till the endpoint is achieved. First the solution turns bluish-green and then attains a reddish-brown color at endpoint.
   - vi. Calculation:

$$\text{normality of FAS} = \frac{A \times N}{V} = \frac{10 \times 0.25}{V} \approx 0.10\ N$$

where A = volume of $K_2Cr_2O_7$ taken, ml=10
N = normality of $K_2Cr_2O_7$ = 0.25
V = ml of FAS used

b. Potassium acid phthalate (KHP) standard:

KHP is used as a standard to estimate the theoretical oxygen demand. KHP reacts with acidified $K_2Cr_2O_7$ as shown in Equation 10.16:

$$\begin{aligned} &2\ KC_8H_5O_4 + 10K_2Cr_2O_7 + 41\ H_2SO_4 \rightarrow \\ &16\ CO_2 + 46\ H_2O + 10\ Cr_2(SO_4)_3 + 11\ K_2SO_4 \end{aligned} \qquad (10.16)$$

Since each molecule of $K_2Cr_2O_7$ has the same oxidizing power as 1.5 molecule of O_2. The equivalent reaction is as follows:

$$2\ KC_8H_5O_4 + 15O_2 + H_2SO_4 \rightarrow 16\ CO_2 + 6\ H_2O + K_2SO_4 \qquad (10.17)$$

Thus, 2 molecules of KHP consume 15 molecules of O_2. So a theoretical oxygen demand can be calculated as:

$$1\text{ mg KHP} \equiv 1.175\text{ mg COD}$$

For standardization, follow the steps given below:

a. Dry an adequate quantity of KHP in an oven at 103°C for 2 h. Cool it to room temperature in a desiccating cabinet.
b. Weigh 0.4252 g KHP and transfer in a 1 L volumetric flask. Dissolve in DDW and dilute up to the mark.

$$0.4252\text{ g KHP/L} \equiv 500\text{ mg COD/L}$$

c. Take 20 ml KHP solution and proceed with as described below for the sample.
d. Calculate COD as described for the sample. If variation in the value is more than ± 10% from theoretical value, prepare fresh solutions of FAS titrant and $K_2Cr_2O_7$ solutions.

8. Procedure

a. Refluxing
   i. Clean the refluxing flasks as described in step 3b.
   ii. Keep a few cleaned and dried glass beads (step 3c) in each flask.
   iii. Add approx. 0.4 g $HgSO_4$ to each flask .
   iv. Add a 20-ml sample or an aliquot diluted to 20 ml with DDW in each flask and swirl.

v. Prepare a flask for blank with 20 ml DDW.

vi. Add exactly 10 ml of 0.25 N $K_2Cr_2O_7$ to each flask and swirl.

vii. Now carefully add 30 ml of the concentrated H_2SO_4 to each flask. The reflux mixture must be thoroughly mixed before heating. Place all the flasks on a hot plate.

viii. Connect the condensers and start cold water circulation. Maintain a smooth flow of water through condenser.

ix. Turn on the hot plate to the highest temperature setting.

x. When the solution starts boiling, set the timer for exactly 2 h.

xi. Turn off the hot plate and allow the samples to cool down to room temperature, keeping all the connections intact.

Note: *(1) The condenser should not be removed before cooling of the refluxing solution to room temperature.*

(2) If any sample after refluxing shows green color, it indicates that all of the dichromate oxidant has been used up. Discard that sample and prepare a new one with more dilution.

xii. Wash the condensers with 40 ml DDW and remove the flasks from the condensers.

xiii. Dilute the refluxing solutions with DDW to make a final volume of 100 ml. Again cool the solution to room temperature.

b. Titration:

i. Add 0.5 ml of ferroin indicator in each flask containing cool refluxing sample / blank.

ii. Titrate with standard FAS.

iii. The endpoint is indicated by a color change from blue-green to reddish-brown.

9. Calculations

$$\text{mg COD/L} = \frac{(A - B) \times N \times 8000}{\text{volume of sample, ml}}$$

where A = volume of FAS used for sample, ml
B = volume of FAS used for blank, ml
N = normality of FAS

10. Range

a. This procedure is used to detect COD in the range of 100–900 mg/L.

b. If sample is expected to have COD higher than this range, it must be diluted.

c. If the sample COD is lower than this range (5–100 mg COD / L) the normality of $K_2Cr_2O_7$ and FAS should be reduced, i.e., 0.025 N and 0.01 N respectively

10.3 Grease/Oil – Partition – Gravimetric Method

Oils, fats, waxes and fatty acids are the major constituents included in this category in domestic wastewater. Industrial wastewater may contain simple esters also. A wide variety of substances, including low to high molecular weight hydrocarbons of mineral origin, gasoline, heavy fuels and lubricating oils, are included in the term "oil." It also represents all glycerides of animal and plant origin that are liquid at ordinary temperature.

Source

Oil and grease are added to wastewater by domestic and industrial activities. Bilge and ballast water, refinery and other industrial plant wastes resulting from the lubrication of machinery and automobile garages add a considerable amount of oil and grease to the water. Rolling mills, gasoline filling stations, slaughterhouses, meat-processing, tanneries, or fat-processing industries are other sources of oil and grease.

Significance

1. Wastewater

The presence of a significant amount of oil and grease in wastewater hinders the transportation of wastes through pipelines. It causes scum (floating sludge) in aeration basins of activated sludge plants, which interferes with the biological oxidation of wastes and produces a low-quality settling sludge.

2. Water stream

If a waste rich in oil and grease content is discharged to the water stream, it coats the gill filaments of fish and causes death. It also destroys algae, plankton and bottom-dwelling organisms. Oil film may interfere with reaeration and photosynthesis and thus disturb the aquatic ecosystem. Hence, waste rich in oil and grease content must not be permitted to discharge without treatment into freshwater streams or the oceans.

Measurement

1. Principle

This procedure involves the extraction of oil and grease with an organic solvent, 1,1,1-trichloroethane. The two immiscible solvents (organic solvent and water) make separate layers. The solvent containing the oil and grease fraction of the wastewater is separated from the aqueous layer. It is dried and evaporated to determine the extractable residue by the Gravimetric method.

Note: *(1) Samples must be collected and preserved as described in Section 3.2.2.1.*

(2) Sometimes high-molecular-weight petroleum hydrocarbons such as crude oils and heavy fuel oils contain nonextractable residues, which cannot be extracted by this procedure. Hence the results will be lower than the exact value.

2. Apparatus

a. Separatory funnel – Two-L capacity with a Teflon stopcock
b. Drying oven – adjusted at 103°C
c. Boiling flask – A flat-bottom boiling flask of 250-ml capacity with standard taper fitting. This can be attached to a distillation unit for solvent recovery.
d. Distillation unit – containing a condenser and a receiving flask of 1-L capacity.
e. Filter Paper – Whatman No. 40 or equivalent with 11-cm diameter.
f. Glass beads – clean and dry.
g. Heating mantle – maintained at 70°C temperature.
h. Desiccating cabinet

3. Reagents

a. H_2SO_4, 50% (v/v) solution: Refer to Section 3.5.3 and Table 3.4.
b. Extracting solvent: Use 1,1,1-trichloroethane, $CCl_3 \cdot CH_3$ as extracting solvent. The molecular weight of solvent is 133.41. It is a clear liquid with a density of 1.32 to 1.34 g/ml and boiling point ranges between 73 to 75°C. This solvent should not leave any residue on evaporation. If it does, distill the solvent before use.
c. Na_2SO_4 : Use anhydrous Na_2SO_4, of analar grade.

4. Sampling

a. Collect 1-L sample in a glass-sampling bottle.
b. Acidify the sample with 5 ml of 50% (v/v) H_2SO_4 solution. Mix well.
c. Take complete 1-L sample for analysis.

5. Standardization

a. Estimation of a standard is not required. The accuracy of the method can be checked with a known sample of vegetable or mineral oil. The 90–95% recoveries will indicate the feasibility of the process.
b. Always run a blank in duplicate using 1L DDW instead of sample and follow the same procedure as described for the sample.

6. Procedure

a. Rinse all glassware with 1,1,1-trichloroethane solvent to remove any traces of oil/grease.

b. Dry the boiling flask and clean glass beads in an oven adjusted at 103°C for 1 h. Cool to room temperature in a desiccating cabinet. Weigh the flask with beads to a constant weight.

c. Transfer 1L sample to a separatory funnel of 2L capacity. Add 5 ml of 50% (v/v) H_2SO_4 at this stage and mix well. pH should be ≤ 2.

Note: If the sample was acidified with H_2SO_4 at the time of collection, omit this step.

d. Rinse the sampling bottle with 20 ml solvent and transfer the washings into the separatory funnel. Repeat this procedure 2–3 times.

e. Shake the separatory funnel vigorously for 2 min.

Note: Release the excess pressure developed during extraction by holding the funnel at an angle and opening the stopcock gently.

f. Allow the solution to stand for about 10 min. for clear separation of the extracting solvent and the water layers. Repeat the extraction procedure and again wait for 10 min for layer separation.

g. Pass the extracting solvent layer into the pre-weighed boiling flask through a funnel having a Whatman paper cone.

h. Rinse the bottle 2–3 times more with 60-ml volume of solvent. Each extraction layer should be transferred to the boiling flask containing the first solvent layer.

i. Connect the boiling flask to the distillation unit and start cold water circulation. Maintain a steady flow of water through condenser.

j. Place the boiling flask on a heating mantle adjusted at 70°C.

k. Evaporate the solvent almost to dryness. Turn off heating mantle and leave the flask on it to cool down. Continue the flow of water through condenser. The last traces of solvent can be removed by using a vacuum for 5 min.

l. Carefully wipe the exterior of the boiling flask with an absorbent clean cloth and a small amount of acetone to remove any traces of water and fingerprints.

m. Place the boiling flask in a desiccator for 1 h. Weigh it immediately. Repeat the process of desiccating and weighing until a constant weight is obtained.

7. Calculation

$$\text{oil/grease, mg/L} = \frac{(A - B) - C}{\text{volume of sample, ml}} \times 1000$$

where A = total weight of flask + beads + residue
B = weight of flask + beads

C = average blank determination, residue obtained from DDW

In the above procedure, 1-L sample volume is taken for the determination of oil/grease content, hence the calculation will be further modified as follows:

$$\text{oil/grease, mg/L} = (A - B) - C$$

8. Important Instructions

Grease or oil may float on the surface, hence it is not feasible to remove a uniform portion of sample for analysis. This will not give an accurate analytical result. Therefore, take complete 1-L sample for analysis.

10.4 Phenols – Amino-antipyrene and Chloroform Extraction Method – Minimum Detection Limit: 1 μg/L Phenol

Organic compounds that contain hydroxyl (OH) group directly bound to a carbon atom of benzene ring are known as phenols. They are acidic in nature because they can release a proton (H^+) on ionization due to the influence of the aromatic benzene ring:

$$C_8H_5OH \rightarrow C_6H_5O^- + H^+ \qquad (10.18)$$

Source

Phenol is recovered from coal tar. It is used as a primary component in the manufacture of synthetic polymers, paints, pigments, resins and pesticides. Hence the effluent of these manufacturing industries is rich in phenols. Considerable amounts of phenol and its substituted compounds are present in the wastes from coal and petroleum industries and refineries.

Significance

1. Toxicity

Industrial effluents containing phenol, when discharged into water streams or sewers, destroy the ecological balance of the streams and interfere with the operation of wastewater-treatment processes. They are toxic to microbial life.

2. Odor and taste

The presence of phenol imparts an objectionable odor and taste to water even at concentrations as low as 0.10 mg/L. It taints fish flesh and imparts an unpleasant odor to fish tissues and other aquatic food.

3. Carcinogenic character

The situation becomes more serious if the water containing phenol is chlorinated. This process will produce chlorinated phenols, which are highly odorous and have objectionable taste. These can be detected even at a concentration of 0.005 mg/L. These are carcinogenic compounds also.

Measurement

1. Principle

The sample is distilled to remove interfering substances such as oxidizing compounds, sulfur compounds and oils and tars. The phenols present in the distilled sample react with 4-amino-antipyrene in presence of potassium ferricyanide [K_3 Fe $(CN)_6$] to form a colored antipyrene dye that is then extracted from the aqueous solution with chloroform. The absorbance of the color is measured at 460 nm.

Since the water samples contain various types of phenolic compounds, the analytical results are expressed as mg/L of phenol (C_6H_5OH).

2. Apparatus

a. Distillation unit – Hach distillation unit as shown in Figure 6.3.
b. UV Spectrophotometer – Any commercially available model set at 460-nm wavelength.
c. Separatory funnel – 500-ml capacity
d. Volumetric flasks – 100-ml capacity
e. pH meter

3. Reagents

Prepare all reagents in DDW free from phenol and chlorine.

a. Phosphoric acid, H_3PO_4, 10% (v/v) solution: Measure 10 ml of 85% H_3PO_4 in a 100-ml volumetric flask. Bring the volume up to the mark with DDW.
b. Copper sulfate: powder
c. Methyl orange indicator: Commercially available solution.
d. Reagents to remove turbidity:
   i.. Sulfuric acid, H_2SO_4, 10% (v/v) solution: Measure 90 ml DDW in a 100-ml volumetric flask. Add carefully 10 ml concentrated H_2SO_4 to it to bring the final volume to 100 ml.
   i. NaCl: Powder, analytical grade.
   ii. Chloroform, $CHCl_3$: Analytical grade.
   iii. NaOH, 2.5N: Refer to Section 3.5.3 and Table 3.5.
e. 4-Amino-antipyrene solution: Weigh 3 g of 4-amino-antipyrene and transfer to a 100-ml volumetric flask. Dissolve in DDW. Make up the final volume to 100 ml with DDW.

Note: Prepare fresh solution daily.

f. Potassium ferricyanide solution, $K_3Fe(CN)_6$: Weigh 4.0 g $K_3Fe(CN)_6$ and transfer to a 100-ml volumetric flask. Dissolve in DDW. Bring the volume up to the mark with DDW.

Note: (1) If solution is turbid, filter it. (2) Store in a brown-glass bottle. (3) Prepare fresh solution weekly.

g. Phenol stock solution: Dissolve 1.0 g of phenol in freshly boiled and cool DDW. Dilute up to 1 L with freshly boiled and cooled DDW.

$$1.0 \text{ ml phenol stock solution} \equiv 1.0 \text{ mg phenol}$$

h. Phenol standard solution:

i. Measure 1.0 ml of stock solution in a 100-ml volumetric flask. Dilute up to the mark with DDW. Mix thoroughly.

ii. Dilute 1.0 ml of above standard solution to 100 ml with DDW.

$$1.0 \text{ ml phenol standard solution} \equiv 0.1\ \mu g \text{ phenol}$$

i. Ammonium hydroxide, NH_4OH, 3M: Measure 30 ml fresh concentrated NH_4OH in a 100-ml volumetric flask. Make the volume up to the mark with DDW.

j. Phosphate buffer: Weigh 10.45 g K_2HPO_4 and 7.23 g KH_2PO_4. Transfer both the chemicals to a 1L volumetric flask. Dissolve the mixture in DDW. Bring the final volume up to the mark with DDW. The pH of this buffer must be 6.8.

4. Distillation of the Sample

a. Set up the distillation apparatus as described in Section 7.5.5. Use a 500-ml Erlenmeyer flask to collect the distillate.

b. Place a stirring bar in the flask.

c. Measure 300 ml of the sample into the distillation flask with a clean graduated cylinder.

d. Add 1.0-ml of methyl orange indicator solution to the flask.

e. Turn on the stirrer for thorough mixing.

f. Add 10% phosphoric acid with the help of a dropper until the color of indicator changes from yellow to orange.

g. Add 0.1 g $CuSO_4$ to the flask and continue stirring to dissolve it.

Note: If the sample is preserved (refer to Table 3.1) at the time of sampling omit steps (f) and (g).

h. Close the distillation flask. Turn the water circulation on and adjust a constant flow of water through the condenser.

i. Turn on the heater and set it to control position 10.

j. Collect about 275 ml of distillate in the receiver. Turn the heater off.

k. Add 25 ml warm DDW to the distillate.

l. Turn the heater on again. Collect another 25 ml of the distillate in the receiver.

m. Turn off the heater and measure 300 ml of the distillate in a flask. This distillate is used for the measurement of phenol content in the sample.

5. Removal of Turbidity

If the distillate is turbid, follow the instructions given below to remove turbidity:

a. Acidify the distillate with 10% H_3PO_4 and distill again repeating steps (4h) to (4m).

b. If a second distillate is still turbid, follow the steps given below:

   i. Take 300 ml distillate in a 1-L Erlenmeyer flask. Add 3–4 drops of methyl orange indicator.

   ii. Add, dropwise, 10% (v/v) H_2SO_4 solution until the color of indicator changes from yellow to orange.

   iii. Transfer the contents of Erlenmeyer flask to a separatory funnel and add 150 g NaCl.

   iv. Add 30 ml chloroform to the separatory funnel and mix thoroughly.

   v. Transfer the chloroform layer to another separatory funnel.

   vi. Repeat steps iv. and v. 3–4 times more with 20-ml volume of $CHCl_3$ in each repetition.

   vii. Now add 3.0 ml of 2.5 N NaOH solution to the second separatory funnel. Mix well. Pour the alkaline layer into a beaker.

   viii. Repeat step vii. 2–3 times more with 2.0-ml volume of 2.5 N NaOH.

   ix. Combine alkaline extracts and keep on a water bath.

   x. Continue heating until $CHCl_3$ has been removed.

   xi. Cool the solution and dilute to 300 ml.

   xii. Distill this solution again as described in Section 4.

6. Standardization

Prepare calibration standards containing 5, 10, 50 and 100-µg/L phenol and standardize as follows:

a. Into four different 100-ml volumetric flasks measure 5, 10 and 50 ml of standard phenol solution containing 100-µg/L phenol.

b. Dilute the solution in each flask up to the mark with DDW. Mix thoroughly.

c. For 100 µg/L concentration take the standard solution directly without dilution.

d. Follow the same procedure of color development as described below for the sample.

e. Prepare a blank with 500-ml DDW instead of standard solution.

f. Plot a calibration curve of absorbance against the concentration of phenol in µg/L present in the standards.

7. Procedure

a. Into two separatory funnels pipette 300 ml DDW and 300 ml sample distillate respectively.

b. Add 2.0 ml of 3M NH_4OH solution to both funnels.

c. Adjust pH 9.5 to 10.0 with dropwise addition of phosphate buffer to each separatory funnel. Stopper the funnels and shake to mix well.

d. Add 1.5-ml 4-amino-antipyrene solution to each funnel. Stopper and shake well to mix.

e. Now add 10.0-ml $K_3Fe(CN)_6$ solution to each funnel. Stopper and shake to mix thoroughly.

f. Wait 2–3 min. for the development of the color. At this stage the solution should be clear and light yellow.

g. Add 30 ml of chloroform to each funnel and stopper.

h. Invert each funnel and temporarily vent. First shake them gently and then vigorously for about 30 seconds.

i. Remove the stoppers. Allow both funnels to stand until the chloroform layer settles.

j. Repeat steps (h) and i. once again to get more extraction with chloroform.

k. Insert a pea-sized cotton plug into the delivery tube of each funnel. The volume of $CHCl_3$ extract should be about 25 ml.

Note: *Filtration through cotton removes the suspended water or particles.*

l. Read absorbance of the sample against blank at 460 nm.

8. Calculation

Read the concentration of phenol in μg/L from the calibration curve or from the direct readout of the instrument.

10.5 Surfactants – Methylene Blue (MBAS) Method – Minimum Detection Limit: 10 μg/L LAS

Surfactant is a substance like detergent, which is added to a liquid to increase its wetting properties by reducing its surface tension. It is also named methylene blue active substance (MBAS) because it is measured by the methylene blue method.

Source

Commonly, linear alkyl sulfonate (LAS) is used as a surfactant in detergents. It comes from mainly the effluent of industries dealing with detergent manufacturing. Another source is domestic water, because different kinds of detergents are used for domestic activities.

Significance

Surfactants are high-molecular-weight macromolecules that are organic in nature and biodegradable. Since they are slightly soluble in water, they cause foaming in streams when discharged along with domestic and industrial effluents. Their presence in water creates foaming in wastewater flow reaching treatment plants also.

Surfactants tend to collect at the air–water interface. During aeration, these compounds collect on the surface of air bubbles and produce very stable foam. This interferes with the activity of aerobic bacteria necessary to carry out the oxidation of organic matter in wastewater. Thus, a poor-quality treated effluent is produced.

Measurement

1. Principle

LAS and other anionic surfactants available in the sample react with methylene blue to produce a blue salt represented as MBAS, according to Equation 10.19:

$$\underset{}{(MB^+)\,Cl^- + RSO_3\,Na \rightarrow} \underset{[MBAS]}{(MB^+)(RSO_3^-)} + NaCl \tag{10.19}$$

Here MB^+ indicates the cation of methylene blue. The blue salt MBAS is then extracted with chloroform. The absorbance of the solution is determined at 652 nm against a reagent blank.

2. Apparatus

a. UV Spectrophotometer – Any commercially available model set at 625-nm wavelength.
b. Separatory funnels – Five of 500-ml capacity with Teflon stopcock.

3. Reagents

a. Chloroform, $CHCl_3$: Use reagent grade.

Note: *Always work with $CHCl_3$ under a fume hood using safety masks, glasses and gloves.*

b. Methylene blue solution: Weigh 0.10-g methylene blue and transfer to a 100-ml volumetric flask. Dissolve in DDW. Bring the volume up to the mark with DDW.
c. Wash solution:
   i. Measure 500 ml DDW in a 1-L volumetric flask.
   ii. Add carefully 13.6 ml concentrated H_2SO_4 to the flask.
   iii. Weigh 100 g hydrated sodium dihydrogen phosphate ($NaH_2PO_4 \cdot H_2O$) and add to the flask.
   iv. Shake the flask for complete dilution.
   v. Dilute up to 1-L calibration mark with DDW.
d. Methylene blue reagent:
   i. Measure 30 ml of methylene blue solution (b) in a 1-L volumetric flask.
   ii. Add 500 ml wash solution to the flask.

iii. Mix well and make up the volume to 1 L with DDW.

e. NaOH, 1 N: Refer to Section 3.5.3 and Table 3.5.

f. H_2SO_4, 1 N: Refer to Section 3.5.3 and Table 3.4.

g. Phenolphthalein indicator: Weigh 0.5 g phenolphthalein disodium salt and transfer to a 100-ml volumetric flask. Dissolve in 50 ml 95% ethyl alcohol. Bring up the volume to 100 ml with DDW.

h. LAS stock solution: Purchase detergent standard solution containing 60 mg/L LAS from Hach Co.

i. LAS standard solution: Measure 1.0-ml of stock LAS solution and dilute it to 100 ml with DDW. Swirl to mix.

$$\text{standard LAS solution} \equiv 0.6\ \mu g\ \text{LAS}$$

Note: *During dilution of LAS add DDW slowly by touching the tip of the pipette to the wall of the flask to prevent foam formation. Swirl the flask gently so no foam will be developed.*

4. Standardization

Prepare calibration standards containing 6, 30, 60, 90 and 120 μg/L LAS and prepare a calibration curve as described below:

a. Rinse five separatory funnels with $CHCl_3$ to remove any sticking LAS residue.

b. Into five different pre-rinsed separatory funnels add 1.0, 5.0, 10, 15 and 20 ml of standard LAS solution.

c. Add corresponding quantities of DDW to make up the final volume to 100-ml in each separatory funnel.

d. Follow the procedure of extraction, color development and measurement for each standard as described below for the sample.

e. Prepare a blank with 100 ml DDW instead of standard solution.

f. Record the absorbance of each standard against blank at 652-nm.

g. Plot a calibration curve of absorbance against the concentration of LAS in μg/L.

5. Procedure

a. Selection of sample volume: Select the sample volume according to the expected LAS concentration as mentioned in Table 10.1.

b. Pre-rinse the separatory funnel with $CHCl_3$ to remove any traces of surfactant.

c. Transfer the selected volume of sample to the separatory funnel carefully without producing foam as mentioned in step (3 i).

d. Add few drops of phenolphthalein solution.

e. Now add dropwise 1 N NaOH solution until a pink color appears.

f. Add dropwise 1 N H_2SO_4 to discharge the pink color.

g. Add 10 ml chloroform and 25 ml methylene blue reagent. Shake the funnel vigorously for 30 sec, then wait for 1 min. for separation of phases.

Note: *(1) Do not allow excessive pressure to develop while shaking the funnel. Remove the pressure by tilting the funnel and releasing the stopcock periodically. (2) Excessive agitation causes emulsion problems. If this has happened, add some drops of isopropyl alcohol to settle the layers. In this condition add isopropyl alcohol to each standard and the blank.*

h. Remove the chloroform layer in another separatory funnel. Rinse the delivery tube of the first separatory funnel with a small quantity of chloroform.

i. Repeat the extraction procedure three times more with 10-ml fraction of $CHCl_3$ each time. Collect all the chloroform extracts in the second separatory funnel.

Note: *If at this stage the blue color in the aqueous phase becomes faint or disappears, discard the sample and repeat the extraction procedure using a smaller volume of diluted sample.*

j. To the combined $CHCl_3$ extracts (collected in the second separatory funnel) add 25 ml DDW and 25 ml wash solution. Shake the funnel vigorously for 30 second. Allow to stand for 1–2 min.

k. Transfer the $CHCl_3$ layer through a small filter funnel with a glass wool plug at the apex into a 100-ml volumetric flask.

l. Add $CHCl_3$ twice in 10-ml fractions each time to extract surfactant from the wash solution. Transfer these extracts to 100-ml volumetric flask also.

m. Rinse the glass wool and the funnel with $CHCl_3$. Collect the washings in the volumetric flask.

n. Place the stopper on the flask and invert it several times for thorough mixing.

o. Determine the absorbance at 652 nm against a reagent blank prepared using DDW instead of sample.

TABLE 10.1
Selection of Sample Volume for the Estimation of LAS

Sample volume, ml	Final volume with DDW, ml	Expected range of LAS concentration, mg/L
500	500	0.01–0.02
400	400	0.02–0.10
250	250	0.10–0.40
100	100	0.40–2.00
20	100	2.00–10.0
2	100	10–100

6. Calculations

Calculate the concentration of MBAS in a sample directly from the calibration curve.

7. Important Instructions

Always collect all the waste chloroform in a distillation flask. Distill it for reuse.

Chapter

Determination of Microbiological Characteristics

Microorganisms are ubiquitous, meaning that they are available everywhere in the environment. They are found in air, land, water, wastewater, food, and even on body surfaces. The different microbes contained in water and wastewater can be divided into two main categories:

1. Non-pathogenic

Microorganisms that are harmless to human beings, i.e., causing no disease, are called non-pathogenic microbes.

2. Pathogenic

Microorganisms that are capable of causing various diseases in human beings are known as pathogenic microbes.

A microbiologist requires some basic equipment and has to follow the specific techniques to isolate, identify and study the impact of non-pathogenic and pathogenic microorganisms present in water and wastewater on human health and on the environment. Such equipment and techniques are described in this section in detail.

11.1. Microbiological Equipment

1. Cultivation Equipment

a. Media

Microbes require an adequate supply of nutrients and a favorable environment for their growth, propagation and survival. A culture medium is used for this purpose. This is a solution containing all the essential nutrients that serve as food and energy sources for the growth of the microbial population.

The culture media can be classified as follows:

i. Liquid medium — prepared by dissolving the salts of essential nutrients in sterilized DDW. It contains no solidifying agent such as agar-agar. It is also known as broth medium.

ii. Semisolid medium — prepared by the addition of 1% agar-agar to the broth medium.

iii. Solid medium — prepared by adding 1.5 to 2.0% agar-agar to the broth medium

Agar-agar is an extract of seaweed, or marine algae. It is a polysaccharide composed of monosaccharide units of galactose. It has no nutritional value. Agar-agar liquefies at 100°C and solidifies at 40°C. Because of this peculiar characteristic it is used as a solidifying agent in the preparation of media. Thus the thermophillic microorganisms requiring higher incubation temperature e.g. 44.5°C can be easily cultivated on solid media with no fear of liquefaction of the medium at this high temperature.

b. Culture Tubes and petri dishes

Glass culture tubes and glass or plastic petri dishes are used to hold culture media for the cultivation of microorganisms. Culture tubes are used to hold broth while petri dishes are used only for solid media. The culture tubes are used for making slants of solid media also.

Before use, culture tubes are sterilized in an autoclave. It is essential to maintain a sterile environment in culture tubes throughout the cultivation of microorganisms. This is achieved either by plugging the tubes with cotton wool or using screw-capped culture tubes.

Petri dishes, which are manufactured with glass or plastic material provide a larger surface area for growth and cultivation. Glass petri dishes can be sterilized several times and reused. Plastic petri dishes are pre-sterilized and disposed of after use.

Petri dishes are available in various sizes, but, for routine analysis, dishes 15 cm in diameter are commonly used. About 15 to 20 ml molten sterile medium is dispensed to sterilized petri dishes and kept for cooling. When the temperature reaches 40°C, the media solidifies. Petri dishes containing solid media are then refrigerated.

Note: During solidification of the medium, water droplets condense on the inside of the cover of the petri dish. Therefore, after inoculation, the petri dish must be incubated in an inverted position so the condensate will not fall on the surface of the solid medium.

2. Sterilization Equipment

Sterilization is a process of rendering a medium or material free of all forms of life. Heat in the form of saturated steam under pressure is the most practical and dependable process for sterilization. Steam under pressure can provide temperatures more than 100°C (the temperature of boiling water). A steam pressure of 15 lb./in^2 maintains temperature at 121°C. If the pressure is increased to 20 lb./in^2, the temperature of sterilization further increases to 126.5°C.

The laboratory apparatus designed to use steam under regulated pressure is known as an autoclave. It is essentially a double-jacketed steam chamber equipped with devices that permit the chamber to be filled with saturated steam. The steam chamber is maintained at a designated temperature and pressure for any required period of time. Thus it can reach a temperature higher than that of boiling water, which helps to kill the organisms present in the solution or in any apparatus.

The autoclave is an essential piece of equipment in every microbiological laboratory. It is used to sterilize various media, dilution water, solutions, glassware used for microbial estimations and discarded cultures and media. Generally, though not always, the autoclave is operated at a pressure of 15 lb./in^2, corresponding to a temperature of 121°C. The time of sterilization depends on the nature of the material being sterilized, the type of container and the volume taken. Culture tubes containing 15–20 ml liquid media can be sterilized in 10–15 min. at 121°C. Large quantities such as 10 L would require an hour or more at the same temperature to ensure sterilization.

3. Inoculating or Transferring Equipment

For propagation and maintenance of active life of microorganisms, cells must be transferred from pure or stock culture to different media or from one culturing container to another. Such transfer is called sub culturing and the technique is called inoculation. Subculturing must be carried out under sterile conditions to prevent possible contamination. The following apparatus are used for inoculation:

a. Inoculation loops and needles

These are made up of inert materials resistant to high temperature, like nichrome or platinum. The loop with needle is inserted into a metal shaft, which serves as a handle. The inoculation needle is easily sterilized by placing it in the blue portion of a Bunsen burner flame till red-hot. When the inoculation loop and needle are not in use, they must be kept in a bottle containing ethyl alcohol.

b. Pipettes

A pipette is another apparatus used for the purpose of inoculation . It is made of either glass or plastic, drawn out to a tip at one end, and with a mouthpiece forming the other end. Use pipettes having graduations distinctly marked and with unbroken tips. Before use, pipettes must be sterilized in an autoclave by keeping them in a canister or they may be wrapped individually in brown paper.

4. Incubating Equipment

The cultivation of microorganisms requires the maintenance of a constant and uniform temperature throughout the incubation period. The instrument used for this purpose is known as an incubator. It resembles an oven and is equipped with a thermostat, which helps to maintain the required optimum temperature for the cultivation of microbes. A thermometer should be kept on each shelf of the incubator. The temperature must be recorded at least twice daily, morning and afternoon.

5. Storing Equipment

A refrigerator is the storage appliance used for a wide variety of purposes such as maintenance and storage of stock cultures between sub-culturing periods, as well as the storage of sterile media to prevent dehydration, and storage of samples and reagents.

11.2 Microbiological Techniques

11.2.1 Cultivation of Microorganisms

All living microorganisms require certain basic nutrients and adequate environmental conditions for their growth, propagation and activity. However, these requirements vary from species to species. In this section, the nutritional and environmental needs for the growth and activity of microbes are discussed.

1. Nutritional Requirements

Nutritional requirements of microorganisms are provided in the laboratory through a variety of media. The following are the essential nutritional elements required by microbes:

a. Carbon: the essential basic element required for the formation of organic molecules such as carbohydrates, protein, nucleic acids, fats etc. These are the fundamental building units of a cell, required for the growth and normal functioning of all cellular structures. Two types of carbon-dependent microbes are known:
   i. Autotrophs — use inorganic carbon in the form of CO_2 for their cultivation. This is supplied by carbon-containing inorganic compounds, used in media.
   ii. Heterotrophs — use carbon supplied by organic compounds used in media. The most common and simple organic compound used in cultivation media is glucose.
b. Energy: Energy is required for the metabolic activities of cellular life. On the basis of energy utilization, microbes can be divided in two types:
   i. Chemotrophs — utilize chemicals either organic or inorganic in nature as their sole energy source, e.g., glucose, hydrogen sulfide (H_2S), sodium nitrate ($NaNO_3$), etc.
   ii. Phototrophs — use radiant energy of the solar system as their sole energy source.
c. Electron donor: An organic or inorganic compound used as a source of electron during oxidation – reduction reactions is known as an Electron donor. Depending on the nature of the electron donor, microbes are classified into two categories:
   i. Lithotrophs — Organisms that use an inorganic compound such as H_2O, H_2S etc., as their source of electron are known as lithotrophs.
   ii. Organotrophs — Organisms that use an organic compound as their source of electron are known as organotrophs.
d. Nitrogen: This is an essential fundamental element of some macromolecules such as proteins and amino acids. These macromolecules are basic units required for the

formation of cellular structures. Nucleic acids, DNA (deoxy-ribonucleic acid) and RNA (ribonucleic acid) also require nitrogen for their synthesis. DNA is the genetic basis of life, while RNA plays an important role in protein synthesis. Some microbes utilize direct atmospheric nitrogen. Some microorganisms use nitrogen supplied in the form of inorganic compounds like ammonium or nitrate salts, while others require nitrogen-containing organic compounds such as amino acids for their growth and survival.

e. Nonmetallic Elements:

   i. Sulfur – The sources of sulfur are either organic compound cysteine–cystine or methionine (amino acids), or inorganic compounds like sulfates, sulfides, and elementary sulfur.

   ii. Phosphorus – This is another element essential for the formation of DNA and RNA and also for the synthesis of high-energy-containing organic compounds like ATP (adenosine triphosphate). It is supplied in the form of phosphate salts in the media.

f. Metallic Elements: Metallic elements like Fe, Ca, Zn, K, Cu, Mn, Mg and Na are essential for varied cellular activities, namely osmoregulation, enzyme activities, and electron transport during bio-oxidation. These nutrients are supplied in the form of different inorganic salts. They are required in micro quantities.

g. Vitamins: Some microbes need vitamins for their growth and activity because vitamins are the cofactors of certain enzymes. Thus, vitamins help to form an active enzymatic system for metabolism.

h. Water: All cells require water in the medium so that low-molecular-weight nutrients can penetrate the cell membrane.

Thus, the three nutritional requirements that are quantitatively most important for the growth and activity of microorganisms are carbon source, energy source and electron donor. So the metabolic capabilities of microorganisms can be broadly classified on the basis of their nutritional requirements, as presented in Table 11.1.

TABLE 11.1
Broad Metabolic Classification of Microorganisms

Source	Nature	Metabolic Group
1. Carbon	• Inorganic • Organic	• Autotroph • Heterotroph
2. Energy	• Chemical • Light	• Chemotroph • Phototropy
3. Electron donor	• Inorganic • Organic	• Lithotroph • Organotroph

2. Environmental factors

In addition to the nutritional requirements, there are three essential environmental factors that influence the growth, activity, and survival of microorganisms. They are (1) the pH of the medium, (2) incubation temperature, and (3) gas requirement. Hence, the successful cultivation of microorganisms requires an appropriate combination of nutrients and optimum environmental factors. For an analyst working in an environmental laboratory it is essential to understand the role of each factor in metabolism.

a. pH of the medium: The pH of the extra-cellular environment greatly influences the enzymatic activities of microbes. The optimum pH for cell metabolism lies in the neutral range for most of the microbes, i.e., 6.5–7.0. A few bacteria can grow at extremes of the pH range. For most species, the minimum and maximum limits fall between 4.0 and 9.0. Fungi and yeast need an acidic environment for activity. Optimum activity occurs at pH between 4 and 6.

Metabolic activities of microbial cells result in the production of waste molecules such as acids from carbohydrate metabolism and alkali from protein degradation. The production of acid or alkali during metabolism may cause change in pH level that can affect the growth and activity of the microbes. To control this change in pH level, some chemicals known as buffers are used in the media. A buffer is a combination of equi-molecular concentrations of a salt of a weak base and a salt of weak acid. The most common buffer used in media is a phosphate buffer containing K_2HPO_4, a salt of weak base and KH_2PO_4, a salt of weak acid.

Under an acidic environment, the K_2HPO_4absorbs excess H^+ ions to form a weakly acidic salt, KH_2PO_4 and a potassium salt with the anion of the strong acid, KCl. Equation 11.1

$$K_2HPO_4 + HCl \rightarrow KH_2PO_4 + KCl \tag{11.1}$$

Under alkaline conditions, KH_2PO_4 releases H^+ ions to form water by combining with excess of OH^- and form a salt of weak base, K_2HPO_4. Equation 11.2

$$KH_2PO_4 + KOH \rightarrow K_2HPO_4 + H_2O \tag{11.2}$$

Thus, a common buffer regulates the pH level required for the cultivation of a specific microorganism. Some other nutrients of media, such as amino acids, proteins, peptones, etc., have natural buffering capability because of their amphoteric nature. Compounds that can act as both an acid and a base are known as amphoteric compounds. They act as natural buffers.

b. Incubation temperature: Microbial growth is greatly influenced by temperature, because the growth processes are all biochemical reactions. The rate of such reactions is directly proportional to the temperature. With increasing temperature, enzyme activity increases, hence influencing the metabolic processes and cell morphology. Thus, temperature will determine the rate of growth and the total amount of cell growth. Each species of microbes grows within a specific range of temperature. On this basis, they are classified into three main categories:

  i. Psychrophiles — These microorganisms are able to grow within a temperature range of – 5°C – 20°C, though they grow best at higher temperatures closer to 15°–20°C.
  ii. Mesophiles — grow best within a temperature range of 25°–40°C.
  iii. Thermophiles — grow best at temperatures between 45° and 60°C. They are further divided into two groups :
    - Facultative thermophiles — grow at 37°C with an optimum growth at 45°C.
    - Obligate thermophiles — grow only at temperatures above 50°C with optimum growth at temperatures above 60°C.

c. Gaseous Requirements: Microorganisms exhibit great diversity in their ability to use free oxygen for cellular respiration. Thus, microbes show different bio-oxidative enzymic activities. They can be divided into the following four main categories:
   i. Aerobic — These bacteria grow in the presence of atmospheric oxygen.
   ii. Anaerobic — These microbes grow in the absence of free atmospheric oxygen.
   iii. Facultative anaerobes — They can grow either in the absence or presence of free atmospheric oxygen. For cellular respiration they use the free oxygen of the environment. If there is a scarcity of O_2 in the environment, they utilize the O_2 present in the combined state such as in nitrates (NO_3^-) and sulfate (SO_4^{2-}).
   iv. Microaerophilic — These microorganisms can grow in the presence of minute quantities of free oxygen.

Table 11.2 represents the nutritional and environmental requirements for the cultivation of microorganisms commonly occurring in water and wastewater.

TABLE 11.2
Nutritional and Environmental Factors Essential for the Cultivation of Microorganisms present in Water and Wastewater

Organisms	Media	pH	Temp. °C	Incubation period (h)	Gaseous requirement
Total plate count	• Nutrient agar	7.0 - 7.1	35 ± 0.5	48 ± 3	aerobic, facultative anaerobes
Total coliforms	• M-Endo agar • Lauryl sulfate broth	7.1 - 7.5	35 ± 0.5	24 ± 2	aerobic, facultative anaerobes
Fecal coliforms	• mFC agar	7.4	44.5 ± 0.2	24 ± 2	Enteric bacteria*
Fecal streptococci	• Azide dextrose	6.6 - 6.8	35 ± 0.5	24 & then 48	Enteric bacteria*
Clostridium perfringens	• Membrane clostridial agar	7.6	37 ± 0.5	24 & then 48	Anaerobic
Salmonella	• Bismuth sulfite		35 ± 0.5	18 - 24	Facultative anaerobes

* *Bacteria pertaining in the intestine are known as enteric bacteria.

11.2.2 Inoculation Technique – Aseptic Method

The transfer of microorganisms from one medium to another is known as subculturing. The knowledge of aseptic technique is essential to a Microbiologist in order to detect the desired microbe present in water and wastewater samples, without any contamination from the microbes present in the surroundings.

Microorganisms are always present in the air and on laboratory surfaces, equipment, and furniture. They can serve as a source of external contamination and interfere with the microbiological growth results. Thus, an appropriate aseptic technique is used during subculturing. This section describes the important steps taken for the aseptic transfer of microbes. The complete procedure is clearly shown in Figure 11.1.

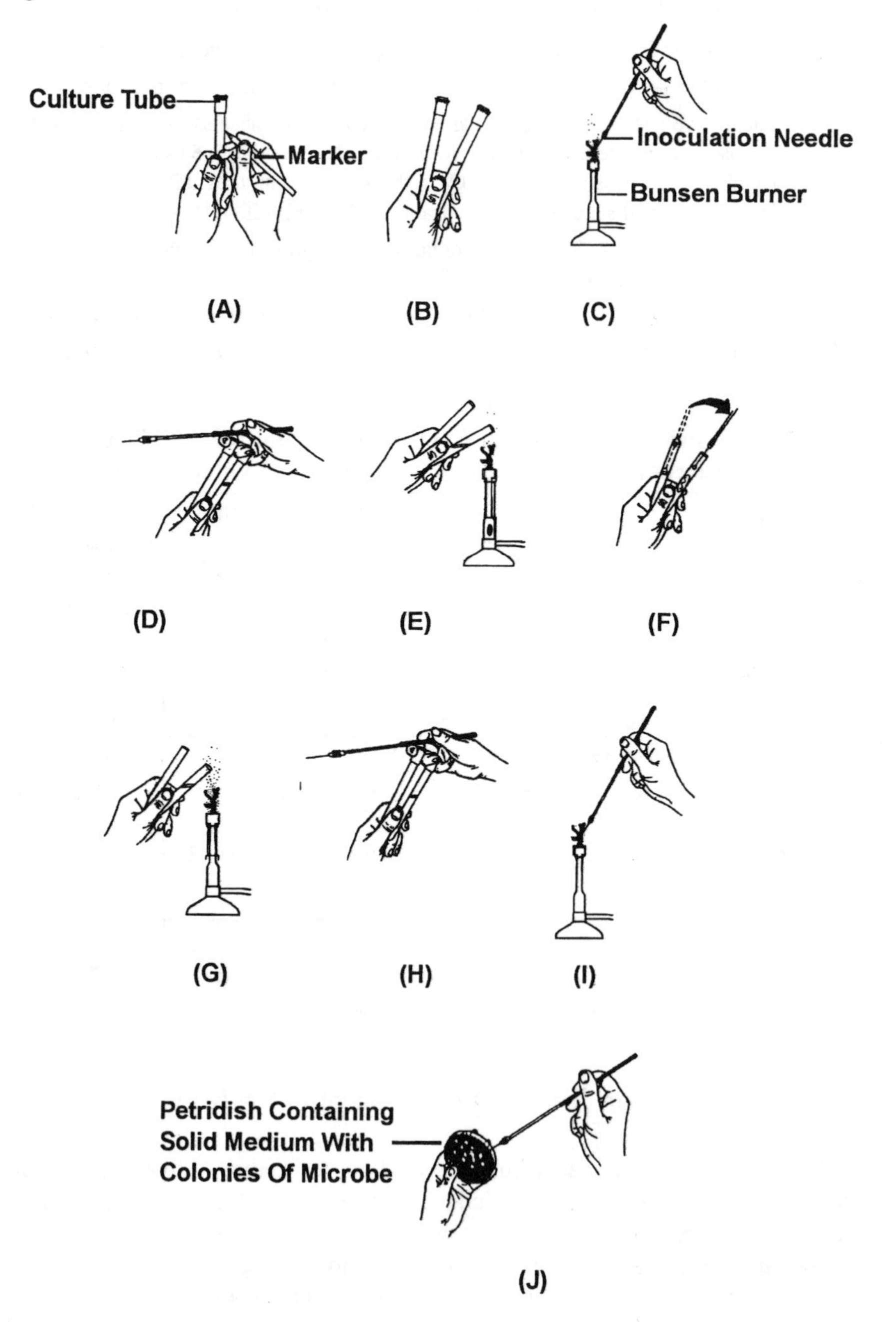

Figure 11.1 - Aseptic technique for transfer of microorganisms

1. Apparatus

a. Bunsen burner
b. Inoculating needle

2. Procedure

See Figure 11.2.

a. Label the culture tube or petri dish containing required medium with the name of organism to be inoculated. The date of inoculation should also be recorded. (A)
b. Hold the culture tube with colonies of stock culture and another tube to be inoculated in your palm so they are separated in a V-shape. (B)
c. Sterilize the loop of inoculating needle by holding it in the blue flame of the Bunsen burner until the entire wire becomes red-hot. (C) Once sterilized, never set the loop down, but hold it in your hand. Allow the loop to cool for 10–20 seconds.
d. Uncap both tubes and hold caps in your hand, holding the sterilized inoculation needle. Never set them down, as there would be a risk of contamination. (D)
e. At each opening and closing steps sterilize the neck of the tubes by passing rapidly through the blue flame of the Bunsen burner. (E)
f. Remove the inoculum from the stock culture tubes with the help of the inoculating loop. Inoculate the broth by slight agitation. Inoculation of slant should be done by drawing the needle upward in a zigzag manner along the surface of the solid medium. (F)

Note: Do not dig the loop into the solid medium.

g. Again sterilize the neck of the tubes by rapidly passing through the blue flame of Bunsen burner. (G)
h. Recap the tubes. (H)
i. Flame the inoculating needle and dip in a bottle containing ethyl alcohol. (I)
j. Incubate the transferred culture at the required temperature.
k. If the culture is to be transferred from a petri dish containing colonies of microorganism growing on solid medium, hold the petri dish near the flame. Touch the sterilized inoculation needle to the surface of the selected colony and transfer to the tube containing broth or slant or to the petri dish containing solid medium. After inoculation, replace the cover of petri dish. (J)
l. After inoculation, incubate the culture tubes at the required temperature. Inoculated petri dishes must be incubated at required temperature in inverted position.

11.2.3 Sample Preparation – Serial Dilution Method

1. Principle

The standard range for the viable count is 30–300 colonies per plate. If the microbial population in water or wastewater sample is not known and a high population is predicted, a series of dilutions must be prepared to obtain a plate count in the standard countable range.

2. Apparatus

a. Petri dishes – with solid media
b. Mechanical pipettes – 1-ml capacity with disposable tips, or sterile glass pipettes
c. Bunsen burner
d. Culture flasks – sterile, eight in number
e. Volumetric flasks – 100-ml capacity

3. Reagents

a. Phosphate buffer stock: Weigh 3.4 g KH_2PO_4 and transfer to a 100-ml volumetric flask. Dissolve in DDW. Adjust pH to 7.2 with dropwise addition of 1 N NaOH. Dilute to 100 ml with DDW. Store this solution in a refrigerator.
b. Magnesium chloride stock solution: Weigh 3.8 g $MgCl_2$ and transfer to a 100-ml volumetric flask. Dissolve in DDW. Bring the volume up to the mark with DDW. Refrigerate.
c. Dilution Water: Prepare any of the following solutions:
   i. Buffered water: Measure 2.5 ml phosphate buffer stock solution and 5.0 ml $MgCl_2$ solution in a 1-L volumetric flask. Bring the volume to 1L with DDW. Sterilize this solution in the autoclave at 121°C for 15 min.
   ii. Peptone water: Weigh 1 g peptone and transfer to a 1-L volumetric flask. Dissolve in DDW and make up to 1 L with DDW. Adjust the pH to 6.8. Sterilize this solution in the autoclave at 121°C for 15 min.

4. Procedure

a. Label the culture flask containing water or wastewater sample as S and the remaining seven culture flasks containing 9-ml water blanks numberically 1–7. Label the petri dishes 1A, 1B, 2A, 2B, 3A and 3B.
b. Transfer aseptically with a mechanical pipette or sterile glass pipette, 1-ml sample from flask S to flask 1 containing water blank. Discard the pipette tip or the glass pipette in the beaker containing sodium hypochlorite solution. The sample has been diluted 10 times. The dilution factor is 10.
c. Mix the contents of flask 1 thoroughly. With a fresh disposable tip or sterile pipette transfer 1 ml from flask 1 to flask 2 containing water blank. Discard the pipette tip or the sterile pipette in the beaker containing sodium hypochlorite solution. The sample has been diluted 100 times. The dilution factor is 100 or 10^2.
d. Repeat the above procedure from flask 2 to 3 and flask 3 to 4. Mix the contents thoroughly. After every transfer, discard the pipette tip or the glass pipette in the beaker containing sodium hypochlorite solution. The sample has been diluted 1,000 times in the former case and 10,000 times in the latter. The dilution factors are 10^3 and 10^4 respectively.

e. Mix well the contents of flask 4. With a fresh disposable tip or sterile glass pipette transfer 0.1 ml of this suspension to Plate 1A. With the same micro-pipette or glass pipette transfer 1 ml suspension from flask 4 to flask 5. Discard the disposable tip or glass pipette in the beaker containing sodium hypochlorite solution. The sample has been diluted 100,000 times. The dilution factor is 10^5.

f. Repeat the same procedure as described in Step e for transfer of suspension from flask 5 to 6 and flask 6 to 7. Table 11.3 presents the level of dilution in each tube and dilution factor corresponding to each transfer.

g. Thus, the required dilution can be obtained by following the dilution procedure described above. The complete serial dilution technique is presented in Fig. 11.2. After inoculation, incubate the petri dishes in an incubator at the required temperature.

TABLE 11.3
Serial Dilution of Water and Wastewater Samples

Tubes	ml of solution present in the tubes	Level of dilution	Dilution factor	ml of solution transferred to plates	Level of dilution	Dilution factor
S	10 ml water or wastewater sample	Nil	Nil	Nil	Nil	Nil
1	9 ml water blank + 1 ml sample from tube S	1	× 10	Nil	Nil	Nil
2	9 ml water blank + 1 ml solution from tube 1	2	× 10^2	Nil	Nil	Nil
3	9 ml water blank + 1 ml solution from tube 2	3	× 10^3	Nil	Nil	Nil
4	9 ml water blank + 1 ml solution from tube 3	4	× 10^4	0.1 ml to plate 1A	5	10^5
5	9 ml water blank + 1 ml solution from tube 4	5	× 10^5	1.0 ml to plate 1B 0.1 ml to plate 2A	5 6	10^5 10^6
6	9 ml water blank + 1 ml solution from tube 5	6	× 10^6	1.0 ml to plate 2B 0.1 ml to plate 3A	6 7	10^6 10^7
7	9 ml water blank + 1 ml solution from tube 6	7	× 10^7	1.0 ml to plate 3B	7	10^7

5. Calculations

The results of the bacterial population are reported as number of colonies / 100 ml of the sample. The following relation will calculate bacterial count:

$$\text{no. of colonies/100 ml} = \frac{\text{no. of colonies} \times \text{df}}{\text{volume of sample, ml}} \times 100$$

where df = dilution factor

a. Plate 1A: Suppose number of colonies on plate 1A = 5

$$\text{Total no. of colonies} = \frac{5 \times 10^5 \times 100}{0.1} = 5 \times 10^8$$

b. Plate 1B: Suppose number of colonies on plate 1B = 20

$$\text{Total no. of colonies} = \frac{20 \times 10^5 \times 100}{1.0} = 20 \times 10^7$$

Similarly, bacterial counts of plates 2A, 2B, 3A and 3B can be calculated.

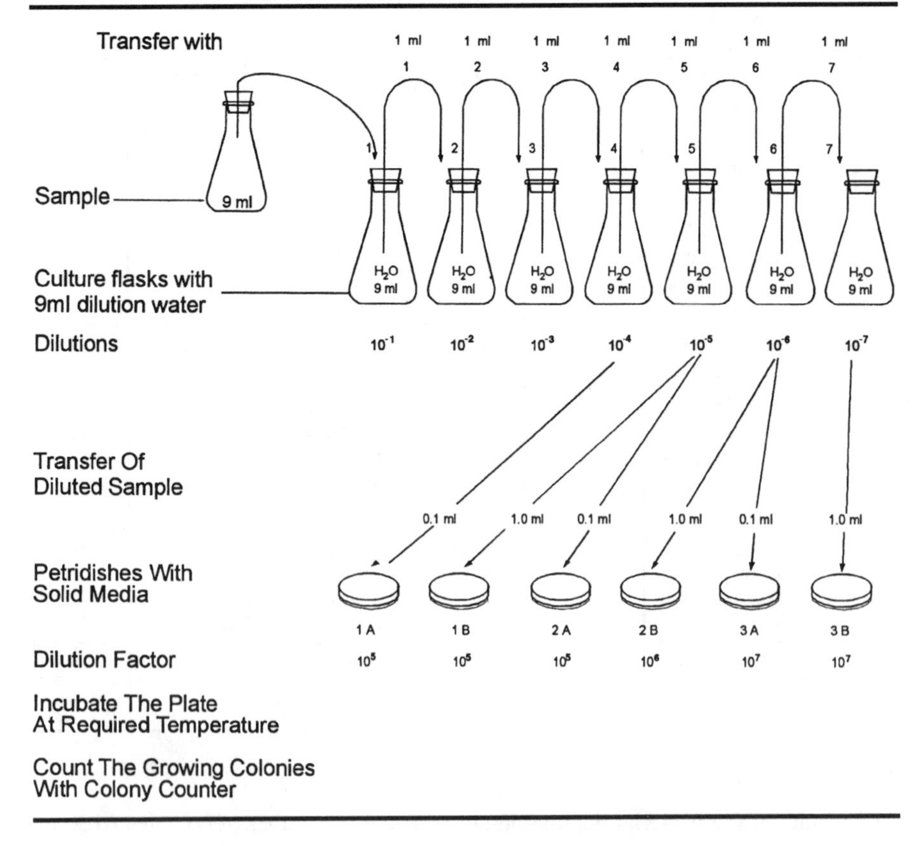

Figure 11.2 - Sample preparation by serial dilution technique

11.2.4 Detection Technique – Membrane Filter Method

Microbes present in water and wastewater can be detected quantitatively by two techniques:

1. Multiple Tube Method

In this technique, different volumes of undiluted or diluted samples are added to a series of tubes containing specific differential broth medium. After incubation, certain tubes show the characteristic changes due to the growth of required organisms. These tubes are considered positive-reaction tubes. Other tubes, which do not show the characteristic changes or growth of microorganisms, are considered negative-reaction tubes.

The most probable number of organisms in 100 ml of the sample can be estimated from the number and distribution of tubes showing a positive reaction.

2. Membrane Filter Method

The membrane-filter technique is more commonly used than the multiple-tube method because:

a. This method is more reproducible and accurate.
b. Results are obtained in a shorter period.
c. Larger volumes of samples can be processed.

This section deals with the schematic representation of membrane-filter technique.

1. Principle

In the membrane-filter method, a known volume of water or wastewater sample is filtered through a membrane filter made up of cellulose esters. The pore size of the membrane-filter is such that it can retain the organisms to be detected on the surface of the membrane. The pore size normally used in membrane filter technique is 0.45-μm.

The membrane filter holding microbes is then placed on selective medium for specific indicator organisms. The medium may be solid agar medium or an absorbent pad saturated with selected medium broth. These plates are then incubated at a specific temperature for the recommended incubation period for different microorganisms.

The microbes retained by membrane will form colonies of characteristic morphology and color, depending on the medium selected. The colonies are then counted with the help of a colony counter and results are expressed as number of colonies per 100 ml of the sample.

2. Apparatus

a. Filtration assembly – A standard commercially available membrane filtration apparatus may be of Pyrex, steel or any non-corrosive and bacteriologically inert material. It consists of a base supporting a porous disc, filter funnel and clamp. The filter funnel may or may not be graduated. Each day the filtration unit including spare funnels and funnel base must be sterilized in an autoclave at 15lb/in^2 pressure. The complete assembly is shown in Figure 11.3.

b. Suction flask – 1-L capacity

c. Vacuum pump

d. Filter membranes – Pre-sterilized membrane filters, 47 mm in diameter with a standard pore size of 0.45-μm, are recommended. For pre-sterilized membranes, the manufacturer should certify that sterilization has neither induced any toxicity nor change in the chemical or physical properties of the membrane. Membranes with grid marks are usually used, as grid marks make the task of colony counting easier. If the membranes are to be sterilized in the laboratory, this can be done either by autoclaving at 115°C

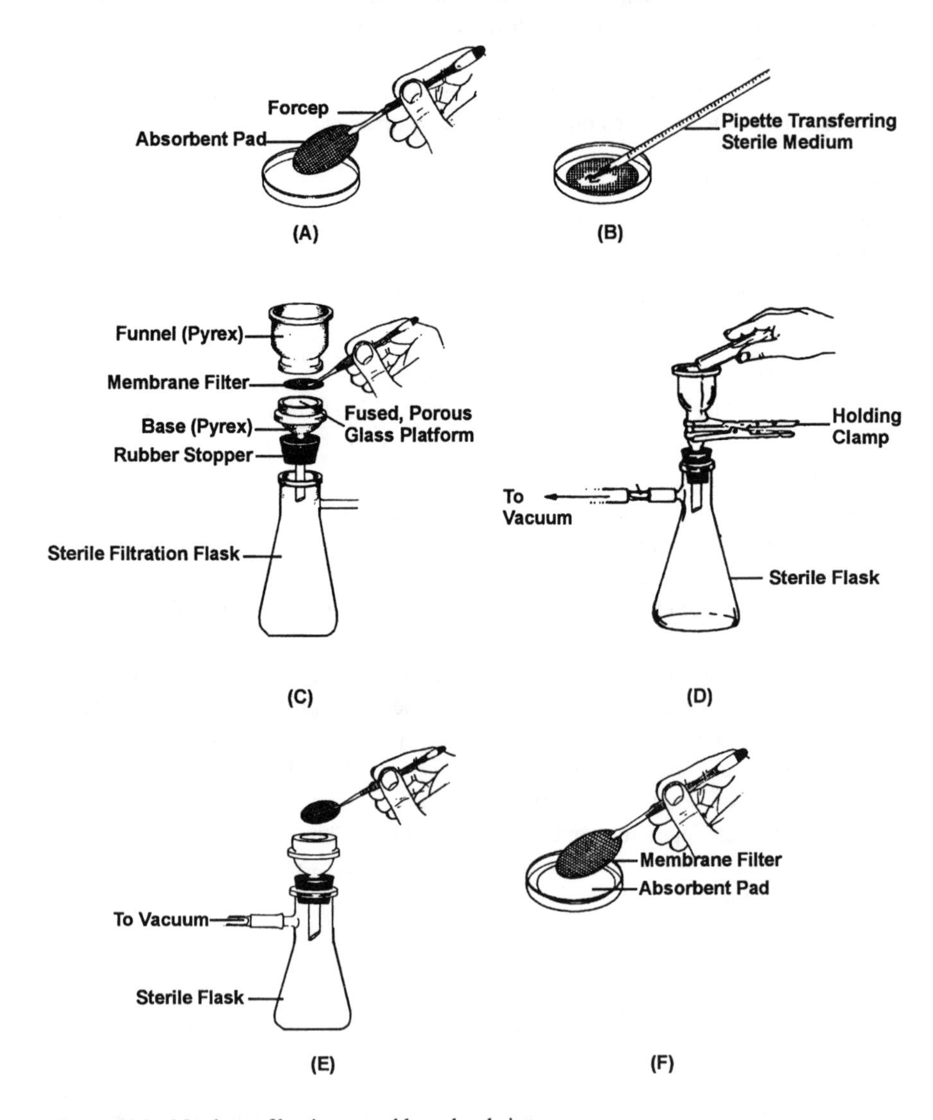

Figure 11.3 - Membrane-filtration assembly and technique

for 10 min. or by boiling gently in DDW for 10 min. Overheating may change the filtration characteristics of membrane filters, so careful temperature control is required during autoclaving. The pre -sterilized membrane filters are preferred.

e. Absorbent pads – Use absorbent pads made up of high-quality paper fibers. They should have uniform absorption capacity and be free from any toxic substance such as sulfite, a microbial-growth inhibitor. Their diameter should be the same as the membrane filters. The thickness should be enough to absorb 2–2.5-ml medium. These are also available pre-sterilized or can be sterilized by autoclaving at 121°C for 20 min. either in containers or wrapped in waterproof paper or foil.

f. Petri dishes – sterilized

g. Forceps – Non-serrated with round or blunt tips, kept in a small beaker containing 95% alcohol

h. Pipettes – Glass pipettes of 10-ml capacity and mechanical pipetting device with disposable tips

i. Incubator

j. Colony counter – equipped with a fluorescent lamp and magnifier

k. Culture media

The details of specific culture media are given with description of individual microorganism in Sections 11.4–11.11.

3. Procedure

See Figure 11.3

a. If dilution of the sample is required, follow the procedure described in Section 11.2.3.

b. Label the petri dishes with the name of the microorganisms to be detected and with required dilution.

c. Using a sterile forceps dipped in 95% alcohol and flamed, transfer aseptically a sterile absorbent pad to each labeled petri dish. (A)

d. With sterile 10-ml glass pipette, aseptically add 2 to 2.5 ml specific medium broth to each petri dish according to the microorganism to be detected. Pour off any excess medium from the saturated pads. (B)

Note: *If petri dishes with solid media are used for inoculation, omit steps (A) and (B).*

e. Aseptically assemble the membrane-filter unit as follows:

    i. Insert the filter base into the neck of 1-L filtration flask.

    ii. Place a sterile membrane filter with the help of a sterile forceps on filter base, grid side up, on the sintered base.

    iii. Place the sterile funnel on the top of the filter disc. Secure the funnel to the filter base by clamping or by means of screw threads. (C)

    iv. Attach the side arm of suction flask to a vacuum pump. Turn on the vacuum pump.

f. Pour the sample aseptically into the funnel and filter it through the membrane filter under vacuum. (D)

g. Wash the inner surface of the funnel with 20 ml sterile DDW when the entire sample has been filtered.

h. Turn off the vacuum pump and remove the funnel. Aseptically remove the membrane with the help of forceps. (E)

i. Transfer the membrane carefully to a petri dish containing either an absorbent pad or medium solidified with agar. (F)

j. Incubate the petri dishes at the temperature and for the time specified for the organism sought.

k. After incubation, count the characteristic colonies on the membrane with the help of a colony counter and record the number of colonies of each organism.

Note: *(1) For different volumes of same sample, the funnel may be reused without boiling, provided that the smallest volumes or largest dilution are filtered first.*

(2) For different samples, use different sterile funnels or if the same funnel is to be used, use it after boiling in a water bath and cooling.

(3) After filtration of each sample, sterilize the complete filtration assembly in the autoclave.

(4) Do not alternate the filtration of known polluted sample with those of treated water samples through the same filtration assembly.

(5) Filter the chlorinated water samples and polluted water samples by separate filtration assemblies.

4. Calculation

$$\text{total colonies}/(100 \text{ ml}) = \frac{\text{colonies counted} \times 100}{\text{volume of sample filtered, ml}}$$

If the sample has been diluted, consider the dilution factor (df) and use the following relation:

$$\text{total colonies}/100 \text{ ml} = \frac{\text{colonies counted} \times \text{df} \times 100}{\text{volume of sample filtered, ml}}$$

11.3 Indicator Microorganisms and their Significance

The bacteria used as indicators of fecal pollution in water are grouped under the heading indicator organisms. They only serve to indicate fecal pollution and therefore can not be taken as a criterion of the quantitative degree of fecal pollution or the presence of pathogenic microorganisms.

For several reasons, monitoring for the presence of specific pathogenic bacteria, viruses and other pathogenic agents in water is impracticable and indeed unnecessary for daily routine control purposes. Pathogenic microorganisms present in water die out more rapidly than the common bacterial species present in human and animal intestines. Although it may be possible to isolate and identify the pathogenic microbes present in contaminated water, especially when it is heavily polluted, but the process involves selective media identification, biochemical, serological, and other identification tests on pure culture. Thus, detection of indicator organisms is an easy and rapid method to confirm fecal contamination.

Bacterial species of known excretal origin belong particularly to the coliform group. *Escherichia coli, fecal streptococci* and *clostridium perfringens* are considered common organisms indicating fecal pollution. These organisms are constantly present in the human intestine. Usually they are present in greater numbers than pathogenic intestinal organisms. The death rate of indicator organisms in water is slower than pathogens. Therefore, it is accepted that whenever pathogenic bacteria e.g., those causing typhoid or paratyphoid, gain access to a water supply through excretal pollution, they are always accompanied by natural organisms that inhabit the intestine, i.e., Fecal coliforms.

The organisms most commonly used as indicators of fecal contamination in water are classified as follows:

1. Primary Indicators: The organisms most commonly used as indicators of fecal contamination are considered primary indicators. The members of the coliform group are included in this category, particularly E. coli, which is undoubtedly fecal in origin.
2. Secondary Indicators: those organisms that are analyzed when the results of coliform tests are positive. This category includes mainly fecal streptococci or *clostridium perfringens*. These microbes occur normally in feces, though in much smaller numbers than E. coli. Thus, the presence of secondary indicators confirms fecal contamination.

Table 11.4 describes the different organisms considered primary and secondary indicators in water analysis to confirm fecal pollution.

11.4 Total Coliform – Membrane-Filter Method

1. Principle

A dilute water or wastewater sample is filtered through a sterile 0.45-μm membrane filter. The filter is then placed on a sterile pad saturated with M-Endo broth or in a petri dish containing M-Endo Agar solid medium. The petri dish is then incubated in inverted position at 35 ± 0.5°C for 22–24 h.

Pinks to dark-red colonies with a metallic surface sheen are counted as total coliform colonies.

TABLE 11.4
Primary and Secondary Indicators of Fecal Contamination in Water

Indicator	Organisms	Characteristics
Primary	Coliform bacteria	Aerobic & facultative anaerobic, Gram–ve, non-spore-forming rods that ferment lactose with the production of acid and gas within 24 ± 2 h to 48 ± 3 h at 37°C, when grown in a medium containing bile salts. The total coliform group includes four genera in the *Enterobacteriaceae* family, *Escherichia, Klesbisella, Citrobactor* and Enterobacter. Out of these genera, Escherichia (E. coli species) appears to be the most common indicator of fecal contamination.
Primary	Fecal coliform	Coliform organisms that are thermotolerant coliforms and ferment lactose at 44.5 ± 0.2°C within 24 ± 2 h.
Primary	Escherichia coli	*E.coli* is one of the species of coliform bacteria that constitute the larger proportion of the normal intestinal flora of humans and other warm-blooded animals in comparison with any other organisms. It is reported that 1g of fresh feces contains about 10^9 *E. coli* organisms. Hence, it is an effective and confirmed indicator of fecal pollution.
Secondary	Fecal streptococci	Gram + ve cocci that form pairs or chains and grow in a medium containing that concentration of sodium azide that is inhibitory to coliform organisms and most other gram–ve bacteria. They grow at a temperature of 45°C. Some species can withstand heating at 60°C for 30 min., grow in a nutrient broth containing 6.5% NaCl and at pH 9.6. Thus these microbes can serve as good secondary indicators.
Secondary	Enterococci	Two strains of fecal streptococci – *S. faecalis* and *S. faecium*, which are normally, present in man and various animals.
Secondary	*Clostridium perfringens*	Gram+ve, anaerobic, spore-forming rods capable of reducing sulfites to sulfides are clostridia bacteria. *Cl. Perfringens* forms a stormy clot in litmus milk medium. It produces spores that survive for a much longer time than the coliform bacteria. Furthermore, these spores are usually resistant to chlorination. This specific characteristic of *Cl. Perfringens* makes it a suitable indicator to study and implement the disinfection procedure. If there is a long gap between the sampling and analysis, this bacterium should be analyzed to ascertain contamination.
Secondary	*Pseudomonas aeruginosa*	Aerobic, Gram-ve, non-sporing rods growing at 42°C. It produces ammonia from the breakdown of acetamide, liquefies gelatin, and hydrolyses casein but not starch. It produces a blue-green pigment, pyocyanin, or the fluorescent pigment fluorescein, or both. These organisms are present in large numbers in sewage, hence can be detected in water in the absence of the immediate source of fecal pollution.

2. Preservation

Samples for bacteriological estimation shall not be preserved. Analysis should be done preferably within 6 h of sample collection. It should be done within no more than 24 h of collection.

3. Apparatus

a. Glassware — sampling bottles, petri dishes, pipettes and pipette containers. Cleaning and sterilize as explained in Section 3.4.2.

b. Volumetric flasks — 1-L capacity.

c. Culture tubes — if dilution of sample is required.

d. Filtration Unit — assemble the complete filtration unit as explained in Section 11.2.4.

e. Colony Counter — colony-counting device with which the plates can be examined by combined reflected and obliquely transmitted artificial light against a dark background. The device with a small fluorescent lamp with magnifier is acceptable.

f. Incubator — set at 35 ± 0.5°C.

g. Disposable bags — to dispose of the used and autoclaved petri dishes and media.

4. Reagents

a. Medium – M-Endo Agar

Tryptose or polypeptone	10.0 g
Thiopeptone or thiotone	5.0 g
Casitone or trypticase	5.0 g
Yeast extract	1.5 g
Lactose	12.5 g
Sodium chloride	5.0 g
Di-potassium hydrogen phosphate	4.375 g
Potassium di-hydrogen phosphate	1.375 g
Sodium lauryl sulfate	0.050 g
Sodium desoxycholate	0.10 g
Sodium sulfite	2.10 g
Basic fuchsin	1.05 g
DDW	1 L
Final pH	7.1 - 7.3

b. Ethyl alcohol (95%) 20 ml

Measure 20 ml 95% ethyl alcohol in a 1-L volumetric flask. Add DDW in the flask and make up the final volume to 1L. Dissolve the above-mentioned ingredients in this DDW containing 20 ml 95% ethyl alcohol. Heat to boiling. If required, solidify this medium by adding 1.5–2.0% agar-agar before boiling and prepare the petri dishes. Cool.

Note: *(1) Do not prolong heating or sterilize by autoclaving. (2) Store the medium in the dark at 2–10°C and discard any unused medium after 96 h.*

5. *Preparation of Sample*

a. Do not open the sterile sample bottle until it is required to collect the sample.

b. Collect sample in a sterile bottle so that water does not flow across your hand and then into the bottle. In a moving stream, point opening of bottle upstream and sweep bottle through water against the current, so there is no hand contamination. The bottle will not be completely filled, so the appropriate mixing of sample can be done.

c. Invert the sample bottle rapidly several times to disperse any sediment.

d. Remove the stopper or cap and retain in the hand.

e. Flame the mouth of the sample bottle and dispense the contents on the membrane filter.

f. If necessary, make dilutions at this stage, as described in Section 11.2.2.

6. *Standardization*

Standardization of the process is not required. However a positive and a negative control should be run to ascertain respectively the quality of the medium and the accuracy of the technique used:

a. Positive Control: Use a pure culture in place of a sample. Transfer aseptically the pure culture cells in a culture tube containing 10 ml sterile dilution water. Follow the same procedure of filtration and incubation as described below for the sample. Make the appropriate dilutions to obtain a range of 20–80 colonies per plate. Set up a positive control for each batch of media prepared. This will indicate the suitability of medium for microbial growth.

b. Negative Control: Use sterile dilution water in place of sample as a negative control. Follow the same procedure as described for positive control for each batch of media and each series of samples. If the plates show any colony, do not record the data from that series of samples. Discard the media. Prepare fresh medium and collect the samples again and reanalyzed. Negative control will indicate the accuracy of the procedure.

7. *Procedure*

a. Follow the procedure of filtration as explained in Section 11.2.4.

b. Place the membrane filters in petri dishes containing pads soaked with M-Endo agar medium or solid M-Endo agar medium without any air entrapment.

c. Invert petri dishes and incubate for 22–24 h at 35 ± 0.5°C.

d. Count the colonies pink to dark-red in color with a specific golden-green metallic surface sheen when viewed with a colony counter.

e. Compute coliform densities from the membrane-filter count within 20–80 coliform colonies per plate. If the number of colonies is more than this range and still countable, count the colonies and calculate the number of total coliform colonies per 100 ml as described below.

8. Calculation

$$\text{total coliform}/100 \text{ ml} = \frac{\text{no. of colonies counted}}{\text{volume of sample filtered}} \times 100$$

If any dilution is done, consider the df and calculate the colonies with the following relation:

$$\text{total coliform}/100 \text{ ml} = \frac{\text{no. of colonies counted} \times \text{df}}{\text{volume of sample filtered}} \times 100$$

9. Disposal

After microbiological examination has been completed, follow the procedure of disposal described below:

a. Autoclave disposable petri dishes at 121°C (15 psi) for 15 min. This will melt the dish and destroy any bacteria growing on it. Remains may then be disposed of in disposal bags.
b. If glass petri dishes are used, the media with bacterial colonies must be autoclaved at 121°C (15 psi) for 15 min and disposed of in disposable bags. The petri dishes must be cleaned and disinfected as described in Section 3.4.2.

11.5 Fecal Coliform – Membrane Filter Method

1. Principle

The water or wastewater sample is filtered through a sterile 0.45-μm membrane filter. The filter is then placed in a petri dish containing a sterile pad saturated with mFC broth or in a petri dish containing mFC agar solid medium. The inoculated dish is then incubated in inverted position at 44.5 ± 0.2°C for 22–24 h.

All prominent blue-colored colonies that appear after incubation are counted as fecal coliform. Non-fecal coliform colonies are gray to cream in color.

2. Preservation

Do not preserve samples for bacteriological estimation. Preferably analysis should be done within 6 h of sample collection, and no more than 24 h.

3. Apparatus

As described in Section 11.4., plus the following:

a. Volumetric flasks – 100-ml capacity
b. Whirl - Pak bags – to wrap the petri dishes
c. Water bath incubator: set at 44.5 ± 0.2°C

4. Reagents

a. Medium – mFC Broth

Constituents:

Tryptose	10.0 g
Polypeptone	5.0 g
Yeast extract	3.0 g
Sodium chloride	5.0 g
Lactose	12.5 g
Bile salt mixture	1.5 g
Aniline blue	0.1 g
DDW	1 L
Final pH	7.4

b. Sodium hydroxide, NaOH 0.2 N: Weigh 0.8 g NaOH and transfer to 100-ml volumetric flask. Dissolve in DDW. Make the final volume to 100 ml with DDW.

c. Rosalic acid reagent, 1%: Weigh 1 rosalic acid and transfer to a 100-ml volumetric flask. Dissolve in 0.2 N NaOH. Make up the volume to 100 ml with 0.2 N NaOH.

Mix the above ingredients in DDW containing 10 ml 1% rosalic acid reagent. Heat to boiling. Do not overheat. After boiling, immediately remove from heat and cool to below 45°C. This medium can be solidified by adding 1.5 to 2.0% agar-agar before boiling.

Note: *(1) Do not sterilize by autoclaving. Rosalic acid reagent decomposes on sterilization in the autoclave.*

(2) Store stock solution in the dark at 2–10°C and discard after 2 weeks or if its color changes from dark red to muddy brown.

For Prepartion of Sample and Standardization, see Section 11.4.

7. Procedure

a. Follow the procedure of membrane filtration described in Section 11.2.4.

b. Place the membrane filters with the help of alcohol-flame forceps in petri dishes containing pads soaked with mFC broth or mFC solid medium with no air entrapment.

c. Keep 4–6 dishes in inverted and horizontal position in a Whirl-Pak bag and wrap properly. Anchor the bag below the water surface to maintain the critical temperature requirement.

d. Incubate in a water-bath incubator for 24 h at 44.5 ± 0.2°C.

e. Count all blue colonies produced by the growth of fecal coliform. Do not count gray or cream-colored colonies, which are due to the growth of non-fecal coliform bacteria.

f. The appropriate range of colonies lies between 20–60 colonies per membrane filter.

g. Report the results as number of fecal coliform colonies per 100 ml.

8. Calculation

$$\text{colonies of fecal coliform/100 ml} = \frac{\text{no. of colonies counted}}{\text{volume of sample filter}} \times 100$$

If the sample is diluted, the above result must be multiplied with df and calculate the colonies with following relation:

$$\text{fecal coliform/100 ml} = \frac{\text{no. of colonies counted} \times \text{df}}{\text{volume of sample filtered}} \times 100$$

9. Disposal

After microbiological examination has been completed, follow the procedure of disposal described below:

a. Autoclave disposable petri dishes at 121°C (15 psi) for 15 min. This will destroy the dish by melting and bacteria growing on it. Remains may then be disposed of in disposal bags.
b. If glass petri dishs are used, the media with bacterial colonies must be autoclaved at 121°C (15 psi) for 15 min and disposed of in disposable bags. The petri dish must be cleaned and disinfected as described in Section 3.4.2.

11.6 Confirmatory Tests for Enteric Bacteria – IMViC Tests

Confirmatory identification of enteric bacilli is highly essential to monitor and control the fecal contamination of food and drinking-water supplies. The members of the Enterobacteriaceae family are quite common in the intestinal tracts of humans and lower mammals. The different genera of this family can be identified and confirmed by the IMViC series of tests representing indole, methyl red, Voges-Proskauer and citrate tests. This section describes the stepwise procedure of these confirmatory tests.

Indole Test

1. Principle

Oxidation of tryptophane to indole by the action of the tryptophanase enzyme is the basic principle of this test. The presence of indole is detected by Kovac's reagent, which produces a cherry-red color.

All the microorganisms do not show this reaction, therefore this test serves as a biochemical marker. Cultures producing red color on addition of Kovac's reagent, are indicated as indole + ve (posititve) and others, as Indole - ve (negative) microbes.

2. Culture

Coliform bacterial colonies isolated from water or wastewater samples.

3. Apparatus

a. Inoculation needle
b. Bunsen burner
c. Culture tubes

4. Reagents:

a. Medium – tryptone broth

Ingredients:

Tryptone (oxoid)	20 g
Sodium chloride	5 g
DDW	1 L
Final pH	7.5

Dissolve the ingredients in DDW and adjust the pH to 7.5. Distribute the broth in 10-ml volumes in culture tubes and autoclave at 115°C for 10 min.

b. Kovac's Reagent:

p - Dimethyl aminobenzyaldehyde	5 g
Amyl alcohol (AR & free from organic base)	75 ml
Hydrochloric acid (concentrated)	25 ml

Dissolve the aldehyde in alcohol. Add the concentrated acid with care to the alcoholic aldehyde solution. Store it at 4°C and protect it from light.

Note: The final color of the reagent should be light yellow to light brown. If the reagent gives a dark color with aldehyde, discard it.

c. Ringer's Reagent:

Sodium chloride	2.25 g
Potassium chloride	0.11 g
Calcium chloride	0.12 g
Sodium bicarbonate	0.05 g
DDW	1 L

Dissolve the ingredients in DDW. Distribute them in 5-ml volumes in culture tubes. Sterilize in the autoclave at 121°C for 15 min.

5. Preparation of culture

a. Pick up two or three colonies of coliform bacteria growing on solid medium with a sterile inoculation needle.

b. Transfer aseptically to a culture tube containing 5-ml Ringer's reagent.

c. Use this bacterial suspension for IMViC tests.

6. *Procedure*

a. Transfer aseptically 0.1 ml of bacterial suspension in a culture tube containing 5-ml portions of tryptose broth medium. In similar manner, inoculate three more culture tubes.

b. Incubate all the inoculated culture tubes at 37°C for 24 ± 2 h.

c. After 24 h of incubation add 0.2–0.3 ml of Kovac's reagent and shake the tubes gently. Record the color change.

7. *Test Confirmation*

Deep red color in the amyl alcohol layer — positive Indole

Retention of original color of the reagent — negative Indole

Methyl Red Test

1. *Principle*

Glucose is the major substrate oxidized by all enteric organisms for energy production. In this biochemical oxidation, different organic acids are produced as end products. The pH indicator methyl red detects the presence of acids. It turns red at pH 4.0, which is an indication of a +ve test. At pH 6.0, though it is an acidic condition, the indicator turns yellow because of lower H^+ concentration. This indicates a -ve test.

For Culture and Apparatus, see Indole Test.

2. *Reagents:*

a. Medium – Glucose Phosphate Broth

Ingredients:

Peptone	5 g
Glucose	5 g
Dipotassium hydrogen phosphate	5 g
DDW	1 L
Final pH	7.5

Dissolve the ingredients in DDW and adjust the final pH to 7.5. Distribute this medium in 10-ml volumes in culture tubes. Sterilize the tubes in the autoclave at 110°C for 10 min.

b. Methyl red reagent: Dissolve 0.1 g methyl red in 300 ml 95% ethyl alcohol and dilute to 500 ml with DDW.

c. Ringer's reagent: see Indole Test.

3. *Preparation of culture: see Indole Test.*

4. *Procedure*

a. Inoculate culture tubes containing 10-ml volumes of glucose phosphate broth as described in Indole Test.
b. Incubate for not less than two days at 37°C or five days at 35°C.
c. After incubation, add 10 drops of methyl red indicator solution to the culture tube and record the color change.

5. *Test Confirmation*

Red color — positive methyl red

Yellow color — negative methyl red

Voges – Proskauer Test

1. *Principle*

Some organisms produce nonacid or neutral end products such as acetyl methyl carbinol from the organic acids, which is detected by alkaline α - naphthol solution (Barritt's reagent).

Development of deep red-rose color in the culture after 15 min following the addition of Barritt's reagent is the indication of + ve test. The absence of red-rose color development is the - ve test.

For culture and apparatus, see Indole Test

2. *Reagents*

a. Medium – Glucose phosphate broth Ingredients: as described in Methyl Red Test.
b. Barritt's reagent – α-naphthol: Weigh 5 g purified α-naphthol and transfer to a 100-ml volumetric flask. Dissolve in absolute ethyl alcohol. Store at 5–10°C. Prepare fresh solution fortnightly.
c. Potassium hydroxide, 40% (w/v): Weigh 40 g KOH and transfer to a 100-ml volumetric flask. Dissolve in DDW. Make up the final volume to 100 ml with DDW.
d. Ringer's reagent: see Indole Test.

3. *Preparation of culture: see Indole Test.*

4. *Procedure*

a. Take two culture tubes inoculated in Methyl Red Test.
b. Incubate for 48 h at 35°C.
c. Add 3 ml α-naphthol solution and 1 ml KOH solution and shake vigorously. Record the color development.

5. *Test Confirmation*

Rose-red color - Positive Voges – Proskauer

No color development - Negative Voges – Proskauer

Citrate Test

1. *Principle*

Some enteric organisms are able to use citrate as their sole carbon source for energy in the absence of glucose or lactose. Citrate is acted on by the enzyme citrase and produces some alkaline products that change the color of Bromothymol blue indicator from green to deep Prussian blue. This is the indication of citrate + ve test. Citrate negative cultures will show no growth There will be no change in the color of the medium.

For culture and apparatus, see Indole Test.

2. *Reagents*

a. Medium – Simmons citrate medium

i. Ingredients:

Sodium chloride	5 g
Magnesium Sulfate	0.2 g
Ammonium dihydrogen phosphate	1 g
Dipotassium hydrogen phosphate$_4$	1 g
Citric acid	2 g
DDW	1 L
Agar-agar	15 g
Final pH	6.8

ii. Sodium hydroxide, 4% (w/v): Weigh 4 g NaOH and transfer to a 100-ml volumetric flask. Dissolve in DDW. Bring the volume up to the mark with DDW.

iii. Bromothymol blue reagent: Weigh 0.4 g bromothymol blue and transfer to a 100-ml volumetric flask. Dissolve in DDW. Bring the volume up to the mark.

Dissolve the ingredients in DDW and adjust the pH to 6.8 with 4% (v/v) NaOH solution. Add 20 ml of bromothymol indicator to each liter of medium. Distribute 10-ml volumes in culture tubes. Sterilize in the autoclave at 115°C for 10 min. Prepare the slants of solid medium.

b. Ringer's reagent: see Indole Test.

3. *Procedure*

a. Using aseptic technique, inoculate the slants of Simmons citrate medium with bacterial suspension by means of stab and streak inoculation.

b. Incubate at 35 ± 0.5°C for 24 to 48 h.

Note: *Stab inoculation: Insert the loop of inoculating needle up to the bottom of solid medium. Streak inoculation: Make streaks on the surface of solid medium with the loop of inoculating needle in different directions. The loop must not dig the solid medium.*

4. Test Confirmation

Growth with Prussian blue color - Positive citrate test

No growth with green color of thymol indicator - Negative citrate test

Table 11.5 presents the significance of IMViC tests for the identification of different species of enteric bacteria.

TABLE 11.5
Significance of IMViC Tests

Microbes	Indole	Methyl red test	Voges - Proskauer	Citrate
Escherichia coli	+	+	-	-
Escherichia aerogenes	-	-	+	+
Klebsiella - Enterobacter group Klebsiella pneumoniae	+ or -	-	+	+
Citrobactor freundii	-	+	-	+
Citrobactor diversus	+	+	-	+
Shigella species		+	-	-

11.7 Fecal Streptococci – Membrane Filter Method

1. Principle

Filtration of a volume of water or wastewater sample through a sterile 0.45-μm membrane filter. Place the membrane filters on an agar medium containing sodium azide and TTC (2,3,5-triphenyl-tetrazolium chloride). Incubate the dishes at 37°C for 4 hr and then at 44–45°C for a further 44 h.
Count all red, maroon or pink colonies as fecal streptococci.
For Preservation and Apparatus, see Section 11.4.

2. Medium: Membrane Enterococcus Agar

a. Ingredients:

Tryptose	20 g
Yeast extract	5 g
Glucose	2 g
Dipotassium hydrogen phosphate	4 g
Sodium azide	0.4 g

DDW	1 L
Agar – agar	15 g
Final pH	7.2

b. Reagent: TTC (2,3,5-triphenyl tetrazolium chloride)1% solution: Weigh 1.0 g TTC in a 100-ml volumetric flask and dissolve in DDW. Make up the final volume to 100 ml with DDW.

Dissolve the ingredients by heating in a boiling water bath. The final pH is 7.2 without adjustment. When the ingredients are completely dissolved, heat for an additional 5 min. Then add 10 ml of TTC reagent. Pour the medium directly into petri dishes without further sterilization. Store poured plates in the dark at 2–4°C. Discard the medium after 30 days.

Note: Sodium Azide is a highly toxic substance if ingested or inhaled, so extreme care must be taken while handling it. Follow the instructions for use and disposal given in Section 2.2.1.

For Sample Preparation and Standardization, see Section 11.4

3. Procedure

a. For membrane filtration procedure, see Section 11.2.4.
b. Keep the membrane filters without any air entrapment on solid glucose azide medium.
c. Incubate the plates at 37°C for 4 h and then at 44 to 45°C for 44 h.

Note: Incubation throughout at 37°C may yield some false-positive results, which do not show the characteristics of fecal streptococci. Incubation at 44 to 45°C will produce selective F. streptococci colonies with fewer false positive results.

d. Count all red, maroon or pink colonies as presumptive fecal streptococci, which reduces TTC to the insoluble red dye formazan. As not all species reduce TTC, pale colonies should not be ignored, especially when maroon or pink colonies are absent or present only in small numbers. Confirm the growth of streptococci by catalase test (see Section 11.8).
e. Colony count should be in the range of standard count per plate.

8. Calculation: see Section 11.4.

9. Disposal

Sodium azide has a tendency to form explosive compounds with metals, so media and solutions containing it should not be discharged through metal pipework or drains. These compounds must be disposed of with extreme care, preferably through plastic pipes. After the tests, media must be decomposed with an excess of concentrated solution of sodium nitrite. Finally, it is disposed of as described in Section 11.4.

11.8 Confirmatory Test for Fecal Streptococci – Catalase Test

1. Principle

Some microorganisms have the ability to degrade hydrogen peroxide by the action of the enzyme catalase:

$$2\ H_2O_2 \xrightarrow{\text{Catalase}} 2\ H_2O + O_2 \qquad (11.3)$$

Bacterium fecal streptococci does not have this capability, i.e., F. streptococci is catalase-negative. The growth of F. streptococci can be distinguished from other gram +ve cocci by the catalase test.

2. Apparatus

a. Screw-cap bottles: 50-ml capacity
b. Bunsen burner
c. Inoculation needle

3. Culture:

Colonies of fecal streptococci growing on glucose azide medium

4. Reagent

a. Ringer's Solution: see Section 11.6, Indole Test.
b. Hydrogen Peroxide solution - 3% (w / v)

5. Procedure

a. Pick some colonies growing in glucose azide medium with the help of an inoculating needle using aseptic technique as described in Section 11.2.2.
b. Unscrew the cap of bottle containing 10-ml Ringer's solution near the flame and dip the needle into it. Shake well.
c. Add few drops of 3% hydrogen peroxide solution and replace the cap.
d. Appearance of bubbles (because of oxygen gas production as shown in equation) indicates the positive catalase activity, which verifies the presence of nonstreptococcal species.
e. The absence of bubbles constitutes a negative catalase test, indicating a probable fecal streptococcus growth.

11.9 Clostridia – Secondary Indicator – Membrane-Filter Method

Clostridium perfringes is an important member of the Clostridia genus. It is associated with fecal contamination, so is thus considered a secondary indicator of such contamination.

1. Principle

After preparation of water/wastewater sample by heat treatment to destroy vegetative bacteria, filter a known volume of sample through a sterile 0.45 µm membrane filter. Place the membrane filters on membrane clostridial agar medium and incubate the dishes anaerobically at 37°C for 24 h and then for 48 h .

Count all the black colonies appearing on the membrane.

2. Preservation: see Section 11.4.

3. Apparatus

All as described in Section 11.4, plus additional apparatus mentioned below:

a. GasPak Anaerobic System: Containing anaerobic jar, hydrogen and carbon dioxide-generating envelopes, room temperature palladium catalyst and anaerobic indicator strips. The complete system is shown in Fig 11.4.

Note: Working of anaerobic jar —

1. *When water is added to the GasPak envelope, hydrogen is produced. The hydrogen reacts with oxygen on the surface of palladium catalyst to form water and develops an anaerobic environment.*
2. *Carbon dioxide is also generated by GasPak generator in sufficient volume to support the growth of microbes in an anaerobic environment.*
3. *An anaerobic indicator strip is a pad saturated with methylene blue solution, which changes from blue to colorless in the absence of oxygen, thus indicating the anaerobic environment in the jar.*

b. Water Bath: adjusted at 75°C.

4. Medium: Membrane Clostridial Agar

a. Basal Medium:

Meat extract	3 g
Peptone	10 g
Glucose	20 g
Sodium chloride	5 g

Agar-agar	15 g
DDW	1 L
Final pH	7.6

Dissolve the ingredients by steaming and adjust the pH to 7.6 with 1 N NaOH. Distribute in 18-ml volumes in screw-cap flasks. Sterilize by autoclaving at 121°C for 15 min.

b. Solution A: Na_2SO_3,10% (w/v): Weigh 10 g anhydrous Na_2SO_3 and transfer to a 100-ml volumetric flask. Dissolve in DDW and bring the final volume to 100 ml with DDW. Sterilize by autoclaving at 121°C for 15 min.

c. Solution B: $FeSO_4$, 8% (w/v): Weigh 8 g crystalline $FeSO_4$ and transfer to a 100-ml volumetric flask. Dissolve in DDW by steaming. Bring the volume to 100 ml with DDW. Sterilize in the autoclave at 121°C for 15 min.

d. Final Solution: Melt 18 ml of the basal medium. Cool to 50°C and add aseptically 1.0 ml of Solution A and 0.1 ml of Solution B. Mix gently and pour carefully into petri dishes.

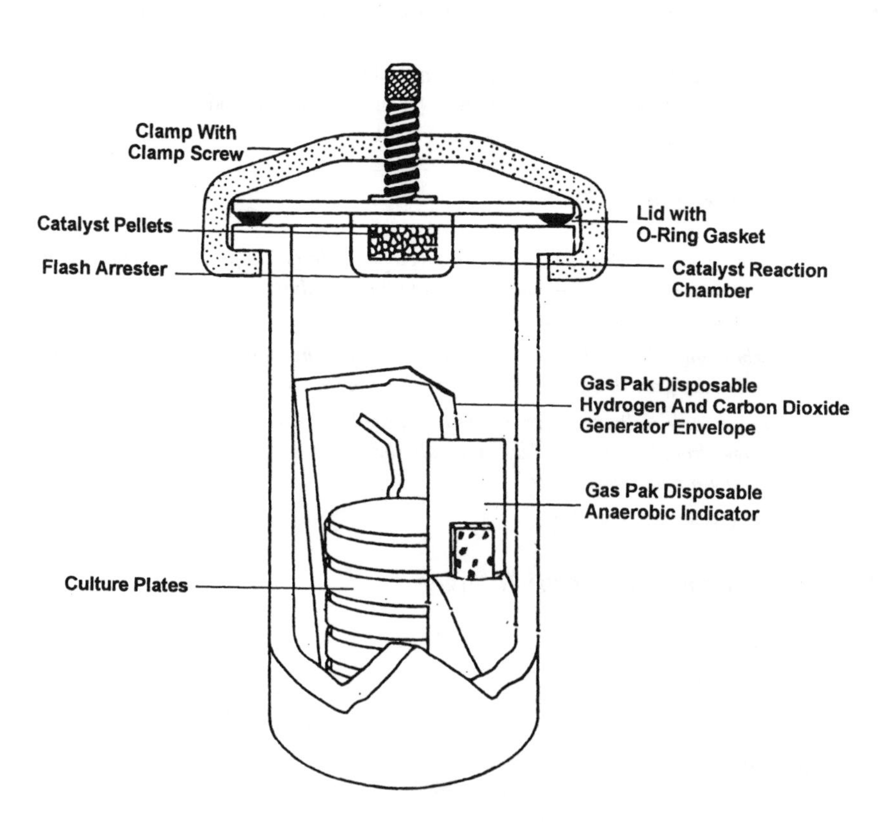

Figure 11.4 Anaerobic jar system for the cultivation of anaerobic bacteria (adapted from Pelczar, M. J., Reid, R. D. and Chan, E. C. S., Microbiology, McGraw-Hill, Inc., 1979, with permission).

5. Sample Preparation

a. Heat the sample to 75°C in a water bath and keep it at this temperature for 5 min.
b. After heat treatment, dilute the sample, if necessary, following the same procedure as described in Section 11.2.3.

6. Standardization: If required, see Section 11.4.

7. Procedure

a. Filter the sample volume following the same procedure as given in Section 11.2.4.
b. Place each membrane facing upward on the surface of a well-dried plate containing membrane clostridial agar medium, without entrapment of air bubbles.
c. Tear off the corner of hydrogen and a carbon dioxide gas-generator envelopes and insert them inside the GasPak anaerobic jar.
d. Place the petri dishes without inversion in the anaerobic jar.
e. Expose the anaerobic indicator strip and place it inside the jar, so that the wick is visible from outside.
f. With the sterile pipette, add 10 ml sterilized water to the gas generator and quickly seal the chamber with its lid.
g. Place the sealed jar in an incubator set at 37°C for 24 to 48 h. After several hours of incubation, the indicator strip changes its color and becomes colorless, indicating the development of an anaerobic environment.
h. Examine the plates after 24 and 48 h.
i. Count all black colonies as sulfite - reducing clostridium.

8. Calculation: see Section 11.4.

11.10 Confirmatory Test for Clostridium perfringes – Litmus Milk Test

1. Principle

Clostridium perfringes has the ability to ferment lactose into lactic acid under anaerobic conditions with the production of CO_2 and H_2 gas. The reaction is catalyzed by β-galactosidase enzyme as presented in Equation 11.4:

$$\begin{array}{c} \text{lactose} \xrightarrow{\beta\text{-Galactosidase}} \text{glucose} + \text{galactose} \\ \downarrow \\ \text{pyruvic acid} \\ \downarrow \\ \text{lactic acid} + \text{butyric acid} + CO_2\uparrow + H_2\uparrow \end{array} \qquad (11.4)$$

The presence of lactic acid is easily detected by change in the color of litmus paper, which is purple at neutral pH and turns to pink under acidic conditions.

2. Culture:

Black colonies of Cl. perfringes growing on the membrane clostridial medium.

3. Medium : Litmus Milk Medium

a. Ingredients:

Skim milk powder	100 g
DDW	1 L

b. Litmus, 10% (w/v) solution: Weigh 10 g litmus and dissolve in 100 ml DDW. Add sufficient volume of 10% litmus solution to skim milk so a bluish-purple color will develop. Distribute in 5-ml volumes in culture tubes. Sterilize in the autoclave at 115°C for 10 min. Before use, the tubes must be steamed to expel any dissolved oxygen and cooled.

4. Procedure

a. Transfer aseptically some colonies growing on membrane clostridial medium to a culture tube containing sterile litmus milk medium.
b. Add to each tube a small nail or a short length of iron wire sterilized by heating to redness immediately before inoculation. This will improve the growth of bacteria in litmus milk medium.
c. Incubate all the culture tubes at 37°C for 48 h.
d. The bottles containing Cl. perfringes show a strong "stormy clot" reaction due to the formation of curd. This is because of the fermentation of lactose. A pink band at the surface of the curd will appear.
e. Due to the production of gas, the curd clot is disrupted, confirming the presence of Cl. perfringes.

11.11 Pathogenic Microorganisms – Salmonella – Membrane-Filter Method

Among pathogenic microorganisms, Escherichia coli, Salmonella, Shigella and enteric viruses are common organisms present in wastewater. For technical and epidemiological reasons, the analysis for pathogenic bacteria has not included in the daily routine a bacteriological examination of water and wastewater. Occasionally, the detection of pathogens is essential when a water supply is suspected of transmitting disease.

This section will cover the detailed procedure to isolate Salmonella in water and wastewater samples.

1. Microbe Description

Salmonellae are facultative anaerobic and gram-negative, motile rods. They are catalase-positive and oxidase-negative. In general, they ferment glucose, mannitol and dulcitol with the production of acid and gas but not lactose or sucrose. They show positive H_2S reactions.

2. Principle

The microorganism is concentrated by filtering a large volume through the membrane-filtration technique. The membrane filter is then placed in a pre-enrichment medium, i.e., buffered peptone water. It is then incubated for 24 ± 2 h at 37°C. After pre-enrichment, the microorganism is enriched by subculturing on an enrichment medium containing tetrathionate, then incubated at 42 ± 1°C for 24 ± 4 h. The growth is then subcultured on selective differential XLD medium and incubated for 18–24 h at 37°C.

Count all red colonies with black centers as salmonella. Usually these colonies look more like black colonies with a red periphery.

3. Apparatus: see Section 11.4.

4. Media

a. Pre-enriched medium—buffered peptone water

Peptone	10 g
Sodium chloride	5 g
Disodium hydrogen phosphate, anhydrous	3.5 g
Potassium di-hydrogen phosphate, anhydrous	1.5 g
DDW	1 L
Final pH	7.2 without adjustment

Dissolve the ingredients in DDW. Distribute in 100-ml volumes in screw-cap wide-mouth bottles. Sterilize in the autoclave at 115°C for 10 min.

b. Enrichment medium—tetrathionate broth

i. Broth base:

Tryptose	7 g
Soya peptone	2.3 g
NaCl	2.3 g
$CaCO_3$	25 g
$Na_2S_2O_3$. 5 H_2O	40.7 g
Ox bile	4.75 g
DDW	1 L

Dissolve all the ingredients except $CaCO_3$ in DDW. After complete dissolution, add $CaCO_3$. Shake thoroughly to dissolve $CaCO_3$. Dispense into bottles and sterilize in the Autoclave at 115°C for 10 min.

ii. Iodine Solution:

Iodine	20 g
KI	25 g
DDW	100 ml

Weigh 25 g KI and transfer into a 100-ml volumetric flask. Dissolve in a little quantity of DDW. Add 20 g iodine to KI solution and shake vigorously. Make up the final volume to 100 ml with DDW.

iii. Brilliant green solution:

Brilliant green	0.1 g
DDW	100 ml

Weigh 0.1 g brilliant green and transfer to a 100-ml volumetric flask. Dissolve in DDW with vigorous shaking. Heat the solution to 100°C for 30 min. and shake constantly to dissolve the dye. After cooling, store in a brown glass bottle in the dark.

iv. Complete medium:

Add exactly 1.9 ml of iodine solution aseptically to each 100 ml of broth base with 0.95 ml brilliant green solution. Mix thoroughly and distribute into sterile containers.

c. Selective Growth Medium - XLD Agar (Xylose Lysine desoxycholate)

i. Basal medium:

Lactose	7.5 g
Sucrose	7.5 g
Xylose	3.75 g
L (-) Lysine hydrochloride	5 g
NaCl	5 g
Yeast extract	3 g
Agar-agar	12 g
DDW	1 L

Reagent: phenol red, 0.4% (w/v): weigh 0.4 g phenol red and transfer to a 100-ml volumetric flask. Dissolve in DDW. Bring the final volume to 100 ml with DDW.

Steam to dissolve above-mentioned ingredients of basal medium in DDW. Add 20 ml phenol reagent solution. Sterilize in the autoclave at 115°C for 10 min.

ii. Solution A

$Na_2S_2O_3\ 5\ H_2O$	34 g
Ferric ammonium citrate	4 g
DDW	100 ml

Place above-mentioned ingredients in a 250-ml Erlenmeyer flask and heat gently to dissolve. Pasteurize at 60°C for 1 h.

iii. Solution B

Sodium deoxycholate	10 g
DDW	100 ml

Place above-mentioned ingredients in a 250-ml Erlenmeyer flask and dissolve. Pasteurize at 60°C for 1 h.

iv. Final medium - final pH 7.3

Melt the basal medium, cool to 50°C and add aseptically 20 ml of Solution A to 1L basal medium. Mix gently. With another sterile pipette add aseptically 25 ml of Solution B. Mix well and pour the final medium into petri dishes.

5. Sample

Preparation as directed in Section 11.4, without dilution.

6. Procedure

The cultivation of salmonella consists of three following stages:

a. Pre-enrichment
   i. Filter the sample by following the procedure of Membrane Filtration described in Section 11.2.4.
   ii. After filtration, place the membrane in wide-mouthed screw-cap flask containing 100 ml of buffered peptone water.
   iii. Incubate the flasks at 37°C for 24 ± 2 h.

b. Enrichment
   i. Transfer 10 ml of culture growing in buffered peptone water to 500-ml screw-cap flask containing 100 ml sterile tetrathionate broth.
   ii. Incubate the flasks at 42 ± 1°C for 24 ± 2 h.

c. Inoculation to selective medium
   i. Inoculate the petri dishes containing XLD agar medium with enrichment culture growing in tetrathionate broth.
   ii. Incubate the XLD plates at 37°C for 18–24 h.
   iii. Examine the plates for characteristic colonies of salmonella.

They form red colonies with black centers or often look like black colonies with a red periphery.

7. Disposal

Follow the same procedure of autoclaving and disposal as described in Section 11.4.

REFERENCES

1. Weiss, G. (Ed.), *Hazardous Chemicals Data Book*, 2nd ed., Noyes Publications, New Jersey, USA, 1986.
2. Tchobanoglous, G. and Eliassen, R., The indirect cycle of water reuse, *Water Wastes Eng.*, 6 (2), 1969.
3. ACS, *American Chemical Society Standards*, 7th ed., American Chemical Society, Washington DC, 1986.
4. Jeffery, G. H., Basset, J., Mendham, J. and Denney, R. C., *Vogel's Text Book of Quantitative Chemical Analysis*, 5th ed., Addison-Wesley Longman Ltd., UK, 1991.
5. Snedecor, G. W. and Cochran, W. G., *Statistical Methods*, The Iowa State University Press, Ames, 1967.
6. Sawyer, C. N., Fertilization of lakes by agricultural and urban drainage, *J. of the New England Water Works Association*, 51, 109, 1947.
7. Pelczar, M. J., Reid, R. D. and Chan, E. C. S., *Microbiology*, 4th ed., McGraw–Hill, Inc., USA, 1979.
8. Lund, H. F., *Industrial Pollution Control Handbook*, McGraw – Hill, Inc., USA, 1972.
9. Pomeroy, R. D., The production of hydrogen sulfide in sewers, Clay Pipe Development Association Ltd., London, 1976.
10. APHA, *Standard Methods for the Examination of Water and Wastewater*, 17th ed., American Public Health Association, 1989.
11. Hatcher, J. T. and Wilcox, L. V., Colorimetric determination of boron using carmine, *Analytical Chemistry*, 22, 567, 1950.
12. Metcalf and Eddy, *Wastewater Engineering: Treatment, Disposal, Reuse*, 3rd ed., McGraw – Hill, Inc., New York, 1991.
13. Ayers, R. S. and Westcot, D. W., Water Quality for Agriculture, FAO Irrigation and Drainage Paper 29, Rev. 1, Food and Agriculture Organization of the United Nations, Rome, 1985.
14. Viessman, W. and Hammer, M. J., *Water Supply and Pollution Control*, 4th ed., Harper and Row, New York, 1985.
15. Jolley, R. L. (Ed.), The environmental impact of water chlorination, Oak Ridge National Laboratory, 1976.
16. Peavy, H. S., Rowe, D. R. and Tchobanoglous, G., *Environmental Engineering*, McGraw – Hill, Inc., New York, 1985.
17. Maier, F. J., *Manual of Water Fluoridation*, McGraw–Hill, Inc., New York, 1963.
18. Fair, G. M., Geyer, J. C. and Okun, D. A., *Water and Wastewater Engineering*, Vol 2, John Wiley & Sons, New York., 1968.
19. Pomeroy, R. D., Parkhurst, J. D., Livingston, J. and Bailey, H. H., Sulfide occurrence and control in sewage collection systems, US EPA, 600 / x-85-052, Cincinnati, Ohio, 1985.
20. Allen, E. R. and Yang, Y., Bio-filtration control of hydrogen sulfide emissions, presented at 84th Annual Meeting of the Air and Waste Management Association, Canada, 1991.
21. Dague, R. R., Fundamentals of odor control, *J. Water Poll. Control Fed*, 44, 583, 1972.
22. Cadena, F. and Peters, R. W., Evaluation of chemical oxidizers for hydrogen sulfide control, *J. Water Poll. Control Fed*, 60, 1259, 1988.

23. Fraser, J. A. L. and Sims, A. F. E., Hydrogen peroxide in municipal, landfill and industrial effluent treatment, *Effluent Water Treatment*, J., 184, 1984.
24. Boon, A. G., Skellet, C. F., Newcombe, S., Jones, J. G. and Forster, C. F., The use of oxygen to treat sewage in a rising main, *Water Pollution Control*, 98, 1977.
25. Henry, J. G. and Gehr, R., Odor control: an operator's guide, *J. Water Poll. Control Fed.*, 52, 2523, 1980.
26. Tomar, M. and Abdullah, T. H. A., Evaluation of chemicals to control the generation of malodorous hydrogen sulfide in wastewater, *Water Research*, 28 (12), 2545, 1994.
27. Anon., Safety Practices for Atomic Absorption Spectrophotometers, *International Laboratory*, 63, International Scientific Communications, Inc., Fairfield, USA, 1974.
28. Maxcy, K. F., Report on relation of nitrate nitrogen concentration in well waters to the occurrence of methemoglobinemia in infants, *Natl. Acad. Sci. Research Council Sanitary, Eng. And Environmental Bull.*, 264, 1950.
29. EPA, Process Design Manual for Nitrogen Control, U.S. Environmental Protection Agency, 1977.
30. Gaudy, A. F. Jr., Biochemical oxygen demand. *Water Pollution Microbiology*, R. Mitchell (ed.), John Wiley & Sons, New York, 1972.

Index

A

B

C

D

H

I

K

L

M

N

O

P

Q

R

S

T